SEPTEMBER 2023



# DRIVERS AND CONSEQUENCES OF TENURE INSECURITY AND MECHANISMS FOR ENHANCING TENURE SECURITY

## A SYNTHESIS OF CGIAR RESEARCH ON TENURE SECURITY (2013–2020)

Rebecca McLain

A Peer-Reviewed Publication
International Food Policy Research Institute
Washington, DC

# INTERNATIONAL FOOD POLICY RESEARCH INSTITUTE

The International Food Policy Research Institute (IFPRI), a CGIAR Research Center established in 1975, provides research-based policy solutions to sustainably reduce poverty and end hunger and malnutrition. IFPRI's strategic research aims to foster a climate-resilient and sustainable food supply; promote healthy diets and nutrition for all; build inclusive and efficient markets, trade systems, and food industries; transform agricultural and rural economies; and strengthen institutions and governance. Gender is integrated in all the Institute's work. Partnerships, communications, capacity strengthening, and data and knowledge management are essential components to translate IFPRI's research from action to impact. The Institute's regional and country programs play a critical role in responding to demand for food policy research and in delivering holistic support for country-led development. IFPRI collaborates with partners around the world.

## AUTHOR

**Rebecca McLain** (mclainrj@pdx.edu) is a Director of Research at the National Policy Consensus Center, Portland State University, Portland, Oregon and a Senior Associate at the Center for International Forestry Research (CIFOR), based in Portland, Oregon.



**Photo credits:** Cover, Miaron Billy, Shutterstock.com

ISBN: 978-0-89629-450-9

DOI: https://doi.org/10.2499/9780896294509

# ACKNOWLEDGMENTS

This publication has been prepared as an output of the CGIAR Research Program on Policies, Institutions, and Markets (PIM), which was active from 2012–2021. PIM was led by IFPRI and supported by the CGIAR Funders. Any opinions expressed here belong to the author(s) and are not necessarily representative of or endorsed by IFPRI or CGIAR.

# ACRONYMS

| | |
|---|---|
| CFE | Community forest enterprise |
| CIFOR | Center for International Forestry Research |
| CLD | Community land demarcation |
| FECOFUN | Federation of Community Forest Users in Nepal |
| FLLC | First-level land certification |
| FLR | Forest landscape restoration |
| IAD | Institutional Analysis and Development |
| IFPRI | International Food Policy Research Institute |
| ILRI | International Livestock Research Institute |
| JVLUP | Joint village land use plan |
| MELA | Monitoring and Evaluation of Land in Africa |
| MSP | Multistakeholder process |
| NGO | Nongovernmental organization |
| PIM | Policies, Institutions, and Markets |
| PIM5 | Policies, Institutions, and Markets Flagship 5 Program on Governance of Natural Resources |
| PPA | Participatory prospective analysis |
| PRM | Participatory rangeland management |
| SDG | Sustainable Development Goal |
| SLLC | Second-level land certification |
| SSA | sub-Saharan Africa |
| VLUP | Village land use planning |

# ABSTRACT

Research since the 1990s highlights the importance of tenure rights for sustainable natural resource management, and for alleviating poverty and enhancing nutrition and food security for the 3.14 billion rural inhabitants of less-developed countries who rely on forests and agriculture for their livelihoods. The specific rights or combination of rights held by an individual, household, or community affects whether they have access to land and resources, as well as how those can be used and for how long. Equally important is the degree to which landholders perceive their tenure to be secure. Landowners are more likely to engage in land and resource conservation if they perceive that the likelihood of losing their land or resource rights is low. Between 2013 and 2021, the CGIAR Research Program on Policies, Institutions, and Markets (PIM) supported researchers to explore the drivers of tenure insecurity and their consequences, as well as mechanisms that can enhance tenure security. Their work focused on rights held by individuals and households, as well as collectively held rights. Studies found that tenure insecurity has a variety of negative consequences for natural resource management, agricultural productivity, and poverty reduction, but the sources of tenure insecurity differ for men and women, and for individual, household, and collective lands. Statutory recognition of customary rights, multistakeholder processes (MSPs) such as for land use planning, and organized social alliances such as Indigenous peoples' groups have emerged as important mechanisms for securing rights or enhancing access to collectively held lands. Long-term partnerships, ongoing engagement, and training for actors at multiple scales increase the likelihood of successful implementation of tenure reforms. Further research on tenure security can contribute to achieving the Sustainable Development Goals, especially by clarifying how customary tenure can provide security and how tenure affects decision-making in multistakeholder platforms.

**Keywords:** Land tenure, women's land rights, tenure security, collective tenure, forest rights

### Highlights

- PIM researchers working primarily in sub-Saharan Africa (SSA) identified urbanization, expanding land markets, population pressure, and agricultural commodification as the primary drivers of tenure insecurity. Across the study countries, women and migrants tended to be disproportionately negatively affected by these factors. Sources of tenure insecurity also differed by gender and, for women, whether they were household heads or spouses of male household heads. These findings underline the need for policymakers to understand intra-community and intrahousehold tenure dimensions when designing interventions.
- Another common finding across cases in SSA was the presence of a substantial percentage of tenure-secure households in areas where customary authorities retain widespread legitimacy. This finding suggests that in contexts where the state lacks land administration capacity, but customary authorities retain legitimacy, strengthening customary or hybrid systems might be a viable alternative to the delivery of tenure security.
- The studies that explored the impacts of tenure (in)security on poverty, food security, and nutrition suggest that greater tenure security tended to be associated with lower levels of poverty and improved food security and nutrition. These relationships held true for agricultural holdings, including women's holdings, collectively managed lands, and

across multiple livelihood sectors linked to natural resources. However, the number of studies examining this topic was relatively small, and more research is needed to better understand the nexus of tenure security, food security, and nutrition.

- Four major obstacles to collective tenure reforms and their implementation were common to most of the studies on collective tenure (in)security, irrespective of the region in which the reforms occurred. The obstacles included (1) resistance or opposition from entrenched powerful interests to changes in the status quo, (2) limited implementation capacity of national and decentralized governments charged with implementing reforms, (3) incompatibilities between social norms and tenure reforms, and (4) resistant or corrupt bureaucratic cultures.
- To address these obstacles, many of the studies highlighted the critical importance of engaging a broad array of social actors when developing and implementing reforms. In several countries, political economy analysis was shown to be a useful tool for identifying the power dynamics at play and motivations of different social actors, and determining who can influence different elements of the tenure and governance system. Broad-based MSPs oriented toward identifying and influencing levers for change were determined to be promising tools for moving from enacting reforms to ensuring that reforms are implemented.
- Policy implications of PIM-supported research on tenure include the need to:
    - Tailor tenure interventions to a range of contexts
    - Adopt "bottom-up" tenure reforms, where feasible
    - Incorporate accompanying measures that enable women to realize their rights
    - Support social movements, user group networks, and MSPs to address power imbalances
    - Use tenure security enhancement as a tool for reducing poverty, and improving food security and nutrition
- Key lessons learned for supporting tenure reforms include:
    - Long-term partnerships are critical to the uptake of reforms
    - Ongoing engagement and training for actors at multiple scales increases the likelihood of successful reform implementation
    - More attention should be paid to strengthening private sector connections to action research
- Research priorities recommended for supporting SDG goals for poverty reduction, food security, and nutrition include:
    - Investigate when, where, and for whom customary tenure systems provide security for lands held by individuals and households
    - Explore how the structure, composition, and processes of multistakeholder platforms influence tenure security, livelihood, and ecological outcomes
    - Clarify how tenure security impacts food security and nutrition
    - Invest in transformational socioecological research and engagement

# CONTENTS

## TABLES



## BOXES

# 1. INTRODUCTION

Since the 1990s, research has highlighted the importance of tenure rights for sustainable natural resource management. Studies demonstrate the role that tenure rights play in alleviating poverty and enhancing food security for the inhabitants of less-developed countries who rely on forests and agriculture for their livelihoods (UNDESA 2019). Tenure consists of "the social relations and institutions governing access to and use of land and resources" (Larson and Springer 2016, 5). It is often conceptualized as encompassing a bundle of rights, including rights of access, use, exclusion, management, and alienation (Schlager and Ostrom 1992).[1] The specific rights or combination of rights held by an individual, household, or community affects whether they have access to land and resources, as well as how those can be used and for how long (Larson and Springer 2016). Equally important, however, is the secureness of tenure rights. Robinson and colleagues (2018, 4) define tenure security as a "landholder's confidence or belief (real or perceived) that agreed-upon rights... will be enforced and upheld by society more broadly." Landowners who perceive that they run little risk of losing their land or resource rights—and therefore that they will benefit from making investments in those resources—are more likely to engage in land and resource conservation (Robinson et al. 2018).

Between 2013 and 2021, the Policies, Institutions, and Markets Flagship 5 Program on Governance of Natural Resources (PIM5) supported applied research that aimed to gain a better understanding of the drivers and consequences of tenure insecurity, as well as how different mechanisms to address insecurity have worked in practice. This research examined tenure security with respect to two major types of rights: those held by individuals or households and those held by collectives. Studies that explored the cross-cutting issue of how tenure security relates to women's rights were integral to both focal areas. The work on tenure security and rights recognition or registration was complemented by research on and engagement with multistakeholder partnerships, many of which seek to influence tenure reforms aimed at strengthening land and resource rights. In Myanmar (Suhardiman, Bright, and Palmano 2019), Lao People's Democratic Republic (Lao PDR) (Suhardiman, Keovilignavong, and Kenney-Lazar 2019), and Peru (Monterroso et al. 2017), political economy analysis provided a lens for comprehending the political dynamics that impede efforts to enact and implement equitable and sustainable tenure reforms. A suite of conceptual, analytical, and training and educational tools were developed to deepen understanding of the consequences of tenure insecurity and provide social actors with the means to promote tenure reforms and ensure their implementation.

PIM-supported research and engagement on tenure security spanned three continents (Africa, Asia, and Latin America) (Table 1). Researchers emphasized collaborative and participatory approaches, engaging partners from multiple

---

1 Schlager and Ostrom divide property rights into two categories: Operational rights and collective choice rights. Operational rights include access and withdrawal rights. They are defined as follows (Schlager and Ostrom 1992, 250): Access is "the right to enter a defined physical property." Withdrawal is "the right to obtain the 'products' of a resource (for example, catch fish, appropriate water, etc.)."
Collective choice rights include management, exclusion, and appropriation rights. They are defined as follows (Schlager and Ostrom 1992, 251): Management is "the right to regulate internal use patterns and transform the resource by making improvements." Exclusion is "the right to determine who will have an access right, and how that right may be transferred." Appropriation is "the right to sell or lease either or both of the above collective choice rights."

sectors and levels of governance. This strengthened research and collective action capacity and built networks for knowledge production in and between the study countries. Research teams prioritized long-term research-action partnerships aimed at influencing policy and social norms to build greater tenure security for rural populations, including traditionally marginalized groups such as women, youth, migrants, and pastoralists.

This report aims to provide an understanding of the key findings from this extensive body of work for policy analysts and advisors with an interest in tenure. Table 1 lists the studies, by country, covered in this publication. Section 1 examines key themes related to the drivers and consequences of tenure insecurity, as well as mechanisms for enhancing rural residents' access to land and resources. We include examples of multistakeholder processes (MSPs) to emphasize their role in advancing tenure security reforms that expand and strengthen rural dwellers' land and resource rights. Section 2 focuses on the tools that PIM-supported researchers have

**TABLE 1 PUBLICATIONS BY COUNTRY**

| Country | Publication(s) |
|---|---|
| Colombia | Arango 2018; Ortiz-Guerrero et al. 2017; Velasquez-Ruiz 2018 |
| Ethiopia | Flintan et al. 2019; Ghebru 2019; Ghebru and Holden 2015; Ghebru and Holden 2019; Quisumbing and Kumar 2014; Ghebru and Girmachew 2020 |
| Ghana | Ghebru, Khan, and Lambrecht 2016; Ghebru and Lambrecht 2017; Lambrecht 2016; Lambrecht and Asare 2016 |
| Guatemala | Gynch et al. 2020; Hellin et al. 2018; Stoian et al. 2018; Stoian and Rodas 2018 |
| Indonesia | Larson et al. 2019; Liswanti et al. 2019; Monterroso et al. 2019b |
| Kenya | Mukasa 2019; Tibalazika et al. 2019 |
| Kyrgyz Republic | Kosec and Shemyakina 2019 |
| Lao People's Democratic Republic | Suhardiman, Keovilignavong, and Kenney-Lazar 2019 |
| Madagascar | Ghebru 2017; Ranjatson et al. 2019 |
| Malawi | Ghebru 2017; Ghebru 2019 |
| Mexico | Gynch et al. 2020 |
| Mozambique | Ghebru, Pitoro, and Woldeyohannes 2015; Ghebru 2019; Ghebru and Girmachew 2019 |
| Myanmar | Suhardiman, Bright, and Palmano 2019 |
| Namibia | Gynch et al. 2020 |
| Nepal | Banjade, Paudel, and Mwangi 2020; Gynch et al. 2020; Pradhan, Meinzen-Dick, and Theis 2019; Sharma et al. 2020 |
| Nigeria | Ghebru 2019; Ghebru and Girmachew 2017; Ghebru and Kennedy 2019 |
| Peru | Larson, Monterroso, and Vigil 2019; Larson et al. 2019; Monterroso et al. 2017; Monterroso et al. 2019a; Monterroso 2019b; Monterroso and Larson 2018a; Monterroso and Larson 2018b; Zamora and Monterroso 2019 |
| Tanzania | Behrman, Billings, and Peterman 2013; Billings, Meinzen-Dick, and Mueller 2014; Dungumaro and Amos 2019; Ghebru 2017; Kisambu et al. 2017; Mwita, Kalenzi, and Flintan 2017; Sulle and Mkama 2019 |
| Tunisia | Frija et al. 2019 |
| Uganda | Doss, Meinzen-Dick, and Bomuhangi 2014; Larson et al. 2019; Monterroso et al. 2019b; Mshale 2019a; Mshale 2019b |

adapted or developed to improve understanding of the drivers and consequences of tenure insecurity, assess tenure reform implementation, and build the capacity of partner organizations to conduct tenure-related research and engagement. In this section, we devote attention to political economy analysis, which warrants special consideration because of its usefulness for identifying likely entry points for tenure reform and reform accountability. Section 3 covers key themes from PIM-supported research and engagement on tenure security, together with implications for policy. Section 4 concludes with key lessons learned on action research intended to support tenure reforms and their implementation, as well as priorities for research that can expand the knowledge base on tenure to advance achievement of the Sustainable Development Goals (SDGs) related to poverty reduction, food security, and nutrition.

# 2. TENURE INSECURITY: DRIVERS, CONSEQUENCES, AND MECHANISMS FOR ENHANCING TENURE SECURITY

To strengthen tenure security, it is important to first understand the drivers and negative consequences of tenure insecurity. The studies conducted under PIM address all of these aspects for both individual and collective holdings, and with specific attention to women's resource rights. This section reviews these research findings, highlighting key themes from each set of studies.

## 2.1 DRIVERS OF TENURE (IN)SECURITY: INDIVIDUAL AND HOUSEHOLD HOLDINGS

Studies examining the drivers of tenure insecurity for individual or household land in Ghana (Ghebru and Lambrecht 2017), Mozambique (Ghebru and Girmachew 2019), and Nigeria (Ghebru and Girmachew 2017) found that the primary drivers of tenure insecurity were urbanization and growing land markets. Other prominent drivers included in-migration (Ghana and Mozambique) and increased economic development (Mozambique and Nigeria). In Ghana, Lambrecht (2016) found the customary norms that limit women's ability to inherit land were a driver of tenure insecurity for women. Indeed, perceptions of tenure insecurity differed by gender and social status, with women and migrants in all three countries being more likely than men or non-migrants to perceive their tenure as insecure. Ghebru's (2019) comparative study in Ethiopia, Malawi, Mozambique, and Nigeria found that women's tenure rights tended to be less secure where demand for land was increasing, such as areas experiencing intense population pressure, land commodification, and agricultural commercialization.

In Uganda, Doss and colleagues (2014) documented internal sources of tenure insecurity, including tense landlord-tenant relationships, clan and family conflicts, and ethnopolitical tensions, as well as external sources, such as government expropriations and increased market demand for land.

The studies revealed important differences within and between households in the sources of tenure insecurity. In Mozambique, women considered collective risks (fear of land loss) to be a greater source of tenure insecurity, whereas men felt that individual tenure risks (fear of land disputes) were the greater source of insecurity (Ghebru and Girmachew 2019). In Nigeria, the perceived risk that the government would expropriate land was higher for women plot managers in male-headed households than for women plot managers in female-headed households (Ghebru and Girmachew 2017). Sources of tenure insecurity in Nigeria also varied depending on migrant status, position in household, proximity to urban centers, and the nature of land markets. Knowing the source of tenure insecurity matters for tenure reform efforts, as tenure interventions and accompanying measures can thus be tailored to solve the right problem.

## 2.2 CONSEQUENCES OF TENURE (IN)SECURITY: INDIVIDUAL AND HOUSEHOLD HOLDINGS

In Tigray (Ethiopia), Ghebru and Holden (2015) found that agricultural productivity was greater for households who both owned[2] and sharecropped land than for those who only sharecropped, with the latter group presumably being less secure.

2 Technically, farmers cannot own land in Ethiopia, since the state claims ownership of all lands. We (and the authors of the study) use the term "owner" to refer to persons holding long-term, heritable leases to cropland.

However, the productivity for sharecropped land varied depending on whether the landlord was a relative, with productivity being higher on plots sharecropped from family members. The landlord's tenure security was even more important for plot productivity. Tenants were less likely to invest labor and inputs in the plots they sharecropped if the owner were female or absent, both of whom tend to have less secure tenure.

Ghebru and Holden (2019) reported that landlords in Tigray were much less likely to be at risk of falling below the poverty line.[3] This is especially consequential for women, since in Tigray, widows and divorced women often must rent out their land. Moreover, individuals with more secure access (such as landlords) were more likely to move out of poverty than tenants. Ghebru and Holden noted that Ethiopia's land laws, which restrict the duration and amount of land that can be sharecropped, may heighten tenure insecurity for landholders who lack the means to farm their land. As a result, there is less land available for tenants, making it difficult for landless and land-poor tenants to move out of poverty through renting or sharecropping land.

In a study covering four regions in Ethiopia, Quisumbing and Kumar (2014) found that all households perceived that possessing a land certificate (obtained through the land registration program described in the next section) increased the likelihood that they would be compensated for their land if it were appropriated and gave them a greater incentive to plant trees. Male-headed households had certificates for a larger percentage of their land than female-headed households. However, because nearly all land in the study area was certified, the magnitude of the difference between men and women in the percentage of their land that was certified was small (between approximately 2 percent for cropped land and 1 percent for all land). Quisumbing and Kumar (2014) noted that the adoption of soil and water conservation technologies was not affected by how much of the household's land was certified, which they attributed to the ubiquitousness of certified land. However, the data showed that female-headed households were less likely to build soil or stone bunds on their land than male-headed households. Quisumbing and Kumar attributed this gender gap in soil conservation investment to women having less knowledge about their land rights. They recommended setting up legal literacy campaigns to increase women's knowledge of their land rights and thus encourage them to make soil conservation investments that would improve their agricultural productivity.

Although limited, there is some evidence that greater tenure security of individual holdings may be positively associated with better health outcomes. Kosec and Shemyakina (2019) examined the impact of land privatization on health outcomes in the Kyrgyz Republic. Using data from national household surveys conducted before, during, and after state and collective farms were privatized, they determined that privatization was associated with better nutrition for children, with children between the ages of one and five years old benefiting the most. After controlling for the availability and amount of land, Kosec and Shemyakina (2019) concluded that "tenure security, the ability to use land as collateral, and all else that comes with having private land" seemed to be associated with the observed improvements in health (Kosec and Shemyakina 2019, 16). However, Kosec and Shemyakina do not offer any insights regarding the mechanism by which tenure security could lead to better nutrition. Instead, they conclude that further research on the links between tenure security and nutrition outcomes is needed to identify how enhanced tenure security, in the form of privatized collective farms, led to improvements in child health in the Kyrgyz Republic.

3 Ghebru and Holden (2019) did not incorporate land certification as a variable in their analysis for this study, possibly because land certification was nearly ubiquitous, as reported by Quisumbing and Kumar (2014) who used the same Tigray dataset.

## 2.3 MECHANISMS FOR SECURING RIGHTS AND ENHANCING TENURE SECURITY: INDIVIDUAL AND HOUSEHOLD HOLDINGS

The majority of PIM-supported research on tenure security for individual and household holdings focused on sub-Saharan Africa (SSA), where a number of countries have adopted rights registration as a mechanism for enhancing tenure security on agricultural land. These land registration programs, which rely on substantial donor funding, are generally designed as "one-size-fits-all" solutions to tenure insecurity, with identical procedures throughout the implementing country. Land registration programs initially aimed to facilitate rural landholders' access to state-owned titles, but since the mid-1990s, attention has focused on documenting and registering existing landholder rights held under customary tenure systems.[4]

PIM-supported studies found considerable variation in the extent to which rights registration programs enhance tenure security in African contexts. Ghebru and Kennedy (2019) reported that urban residents in Nigeria were willing to pay for land registration (such as a certificate of occupancy), whereas rural residents felt that customary institutions provided adequate tenure security. They attributed the lack of interest in land registration to a centralized land administration system that is difficult for rural residents to access, along with a registration process that is overly complex, time-consuming, and subject to corruption. Moreover, the national government lacks the capacity to maintain the land registration system, calling into question the ability of rights registration to deliver tenure security over the long term. Given the weaknesses of the existing land administration system, Ghebru and Kennedy (2019) offer several recommendations for improving the system, including (1) expanding public awareness of land administration procedures, (2) increasing transparency of land registration processes and reducing the time required to process registration demands, (3) replacing paper-based systems with digitized records, (4) enhancing the human and financial capacity of land administration services, and (5) prioritizing registration initiatives in places where demand for certificates of occupancy is high, such as peri-urban areas.

Land rights certification has been more successful in Ethiopia's highlands, where a low-cost, low-tech[5] land rights registration process, known as first-level land certification (FLLC), was introduced in the late 1990s and early 2000s. The system, which provides long-term and heritable use rights to land, has successfully increased tenure security for many landholders (Deininger et al. 2008). The extremely high level of tenure insecurity in Ethiopia's highlands prior to the intervention was likely a major factor in this success. Second-level land certification (SLLC), introduced in 2013, registers the same use rights as FLLC but uses digital technology to map parcels and register rights. SLLC had no impact on tenure security overall, but more land appeared to be available in SLLC areas, as reflected by a decrease in distress rentals and an increase in sharecropping (Ghebru and Girmachew 2020).[6] As in Nigeria, perceptions of tenure security were positively associated with possessing an SLLC in more urbanized areas, but were not affected in rural areas. These findings suggest that SLLC programs should target peri-urban areas, where positive effects are more likely (Ghebru and Girmachew 2020).

Research by Ghebru (2019) found that policy reforms aimed at increasing tenure security for women in Ethiopia, Malawi, Mozambique, and

---

4 Chimhowu (2019, 898) defines customary tenure as "collectively owned land usually under the authority of traditional leadership."

5 Use of traditional methods for demarcating land and paper-based registration.

6 Gebregziabher and Holden (2011, 48) define distress sales and rentals as "a situation where a property is sold or rented due to an urgent need (e.g., need for cash to buy food or to cover medical expenses), and often this sale or rental happens at a reduced price due to the weak bargaining power of the person selling or renting out."

Nigeria have been weakly or unevenly implemented. The lack of effective implementation has reduced the tenure security of women, migrants, and poor landholders, all groups that are more likely to have subsidiary and undocumented land rights. Factors contributing to limited reform implementation include inadequate financial and technical capacity on the part of local and national governments, as well as rent-seeking and corruption by elites and government officials. To reduce the risk of ineffective implementation, Ghebru (2019) calls for policy reforms to be more closely aligned with local practices and administrative capacities, and to improve women's ability to assert their rights to land. These recommendations include (1) land registration processes that are more participatory, as in Ethiopia, (2) community-based initiatives to expand women's legal literacy, (3) engagement of women in drafting land legislation, and (4) explicit direction to incorporate joint registration and rights documentation in land registration processes.

Evidence suggests that the notion of separate state and customary tenure systems operating in parallel, rather than interdependently, fundamentally misrepresents the nature of legal pluralism in Africa (Boone 2019; Boué and Colin 2018; Chimhowu 2019). Instead, most African tenure systems are better conceptualized as neo-customary or hybrid systems that incorporate elements of state and customary approaches for allocating and securing tenure rights to land and resources (Chimhowu 2019). Some scholars have suggested that supporting these hybrid systems, rather than imposing costly and ineffective Western-inspired land registration systems, might prove more effective at delivering tenure security to rural residents (Boué and Colin 2018; Valkonen 2018).

PIM-supported researchers documented the presence of hybrid state-customary tenure systems in Madagascar (Ranjatson et al. 2019). Like many African countries, Madagascar has enacted land reforms that encourage smallholders to obtain land certificates, which can be applied for through local land offices. Although these land certificates are much cheaper and less time-consuming to obtain than state-issued titles, the costs are still beyond the financial capacity of most rural inhabitants. Meanwhile, communities have developed hybrid tenure systems that blend elements of customary and state systems, and appear to provide adequate tenure security for many rural landholders (Boué and Colin 2018; Valkonen 2018). Ranjatson and colleagues (2019) suggest investigating locally developed hybrid systems to identify their weaknesses with respect to delivering tenure security. Once those weaknesses are identified, appropriate tenure or other policy and legal reforms can be encouraged, as well as relevant accompanying measures.

Ghebru and colleagues (2016) reached a similar conclusion in Ghana, where neither the state nor the customary systems delivered adequate tenure security to rural landholders. They described the state system as largely nonfunctional, yet their data showed that smallholders, particularly women and migrants, perceived customary rights to be less secure than market-based rights. They recommended targeting reforms at the customary systems, such as capacity-building of the customary land secretariats, in ways that would enhance tenure security for women and migrants. They also suggested limiting land rights certification to areas where land markets are very active and demand for land certificates is likely to be strong. Lambrecht and Asare (2016) concurred, but cautioned that developers of reform programs need to recognize the considerable heterogeneity within Ghana's customary tenure systems. They emphasized the importance of developing more in-depth knowledge of disparate tenure systems, so that tenure reforms can take those differences into account.

Legal aid programs are another mechanism to improve tenure security for individual and household holdings, and are often implemented as an accompanying measure to land registration programs. Legal aid encompasses a variety of approaches, such as establishing legal clinics that are readily accessible to community members, providing community members with legal training (such as community paralegals), offering legal awareness training for vulnerable groups, assisting with dispute resolution, or providing free legal advice and financial assistance to women so that they can assert their rights (Behrman, Billings, and Peterman 2013; Billings, Meinzen-Dick, and Mueller 2014). Legal aid programs vary in their focus, with some structured specifically to assist women, while others provide legal assistance for the broader community (Billings, Meinzen-Dick, and Mueller 2014). In 2013, Tanzanian lawmakers drafted legislation to establish a regulatory authority that would license legal aid providers and monitor the provision of legal aid services, essentially professionalizing what had previously been an informal service (Billings, Meinzen-Dick, and Mueller 2014). In a case study that drew upon qualitative research on legal aid and women's land rights in Uganda and Tanzania (Behrman, Billings, and Peterman 2013), Billings and colleagues (2014) found that the high level of educational and training requirements in the proposed legal aid bill was likely to significantly reduce the pool of eligible applicants from rural communities, particularly for women, who tend to have less formal education than men. They concluded that legal aid programs should use appropriate standards for education and training to make it feasible to recruit paralegals, particularly women, in remote areas.

## 2.4 DRIVERS AND CONSEQUENCES OF TENURE INSECURITY: COLLECTIVE HOLDINGS

Several studies examined the drivers and consequences of tenure insecurity on lands for which collective rights—either ownership or long-term use and/or management rights—had been recognized or devolved to communities. Among others, relevant research-engagement projects were implemented in Colombia (Arango 2018; Velasquez-Ruiz 2018), Indonesia (Liswanti et al. 2019), Kenya (Mukasa 2019), Nepal (Banjade, Paudel, and Mwangi 2020), Peru (Monterroso and Larson 2018a; 2018b), Tanzania (Dungamaro and Amos 2019; Kisambu et al. 2017), and Uganda (Mshale 2019a; 2019b). Other relevant studies included a comparative case study of conflict management in Indonesia, Peru, and Uganda (Larson, Monterroso, and Vigil 2019); case studies documenting the outcomes of community forest enterprise (CFE) development in Guatemala (Stoian et al. 2018) and Nepal (Sharma et al. 2020); a study of Mozambique's community land demarcation (CLD) program (Ghebru, Pitoro, and Woldeyohannes 2015); and a comparative case study of CFEs in Guatemala, Mexico, and Nepal (Gynch et al. 2020). Key themes from this body of work include the following:

**Lack of titles/rights recognition:** The lack of statutory recognition of customary or collective rights to land was a source of tenure insecurity for collectively held lands in Colombia (Arango 2018; Velasquez-Ruiz 2018), Indonesia (Liswanti et al. 2019), Mozambique (Ghebru, Pitoro, and Woldeyohannes 2015), Peru (Monterroso et al. 2017), and Uganda (Mshale et al. 2019b). Without such recognition, communities are in a weak legal position if the national government issues a logging or mining concession for the land in question, a situation that arises with some frequency in these countries.

**Weak law enforcement:** Weak enforcement of laws and policies by the state was a source of tenure insecurity in Uganda (Mshale et al. 2019a), Peru (Larson et al. 2019), and Tanzania (Kisambu et al. 2017). In participatory prospective analysis (PPA) workshops in Uganda, participants indicated that weak enforcement led to land disputes between community members and more powerful outsiders (Mshale et al. 2019a; 2019b).[7] Likewise, focus groups in Peru reported lack of enforcement by government officials to be a source of insecurity in cases of

7 Participatory prospective analysis is a scenario-building activity that brings together multiple stakeholders to identify driving forces affecting the topic of interest (in this case, forest tenure), develop a set of possible futures, and then find possible pathways for achieving a desired future.

encroachment by outsiders, leading to conflicts between villagers and outsiders (Larson, Monterroso, and Vigil 2019). In Tanzania, a key issue was the lack of enforcement of the legal requirement to include grazing areas and livestock corridors in village land use planning (VLUP), a situation that resulted in mistrust between pastoralists and village leaders (Kisambu et al. 2017).

**Lack of effectiveness of community land management committees:** In Guatemala, Mexico, Namibia, and Nepal, limited management capacity and skills on the part of community land management committees were a source of tenure insecurity in some areas (Gynch et al. 2020). Nongovernmental organizations (NGOs) have played a major role in addressing the capacity and skills gaps of such committees. However, over the long run, reliance on NGOs may pose a threat to the system's sustainability if the NGOs shift their attention to other priorities or their funding sources fluctuate (Ghebru, Pitoro, and Woldeyohannes 2015; Gynch et al. 2020). Gynch and colleagues (2020) found that in Namibia and Mexico, the establishment of joint ventures involving community forest institutions and private enterprises has enabled some community forest institutions to overcome their management weaknesses and finance themselves. In both countries, however, the national governments have provided supportive enabling environments that allowed the CFEs to make a profit from managing their natural resources.

**Lack of awareness of rights:** Community and user group members' limited awareness of their statutory rights to land is a source of tenure insecurity for collectively held lands in some countries. For pastoralists in Tanzania, lack of knowledge about land rights put them at a disadvantage against agriculturalists who encroached upon pastures and livestock watering sites (Dungumaro and Amos 2019). In Uganda, participants in PPA workshops mentioned that the lack of knowledge of their rights was a key force driving community members' inability to actualize those rights (Mshale et al. 2019a, 2019b).

**Lack of coordination across levels of governance and sectors:** For collectively held lands, tenure insecurity was often linked to inconsistencies in government policies between sectors and between governance levels. Both of these contributed to a generalized lack of certainty among community members about what was legally permitted in community forests. Uncertainty regarding the distribution of authorities and responsibilities at the subnational level was common in countries undergoing decentralization, leading to inconsistent implementation of policies and tensions between national and subnational governance entities (see also Mwita, Kalenzi, and Flintan 2017; Monterroso and Larson 2018a).

**Economic development and financial policies:** In both rangeland and forest ecosystems, economic development policies that provide incentives for converting these ecosystems to agricultural land were a source of tenure insecurity for collectively held lands. In dryland areas of Tanzania (Dungamaro and Amos 2019) and Tunisia (Frija et al. 2019), such policies reduce access to grazing lands and watering sites, threatening the viability of pastoralist and agropastoralist livelihoods. In forested areas, such as the Peruvian Amazon (Monterroso et al. 2017) and Colombia (Ortiz-Guerrero et al. 2017), policies favoring agricultural development provide incentives for residents and outsiders alike to convert forested land to cropland, reducing community members' access to forest products and contributing to environmental degradation.

**Violent conflicts and land conflicts among user groups or communities:** In Colombia, tenure insecurity for collectively held lands is a side effect of the violent conflict linked to the decades-long drug wars. Arango's (2018) analysis of collective land tenure in Colombia found that nearly half of the

Indigenous *resguardos*[8] are situated in municipalities with the highest displacement rates linked to conflict. Following Colombia's peace agreement, the national government has sought to implement territorial land-use planning in areas previously controlled by the Revolutionary Armed Forces of Colombia. However, Indigenous and Afro-descendant authorities, who legally must be consulted, have often been left out of these planning processes, placing their land rights at risk (Arango 2018; Velasquez-Ruiz 2018). In Tanzania, an effort in one village to implement joint village land use planning (JVLUP), which would have provided the village with a certificate of customary rights occupancy, was stymied by conflict between pastoralists and agriculturalists over how much land should be set aside for grazing (Mwita, Kalenzi, and Flintan 2017).

**Beliefs and social norms:** Beliefs and social norms within communities are a source of tenure insecurity for women (Monterroso et al. 2019b), Indigenous peoples (Zamora and Monterroso 2019), and pastoralists (Sulle and Mkama 2019). In Peru, participants in PPA workshops believed that neither the national nor regional government places a priority on Indigenous issues, an attitude which they felt was a contributing factor to the delays in implementing native community titling legislation (Monterosso et al. 2019a). In East Africa, pastoralists are often viewed by many agriculturalists with disdain, and by government officials as insufficiently modern. They rarely participate in village and district meetings, in part because of their mobility patterns, but also because meetings are often scheduled on market days when they are selling their livestock, or because they are intentionally ignored (Sulle and Mkama 2019). Moreover, women and pastoralists often are not well-represented in community forest management committees, and even if physically present, they may not feel safe sharing their opinion about management decisions (Sulle and Mkama 2019).

## 2.5 MECHANISMS FOR SECURING RIGHTS AND ENHANCING ACCESS: COLLECTIVELY HELD LANDS

Since the 1990s, various tenure and institutional reforms aimed at reducing tenure insecurity or increasing access to collectively held land and natural resources have been implemented in Africa, Asia, and Latin America. Table 2 lists the collective tenure reforms studied by PIM-supported researchers. Alongside tenure reforms, MSPs have become increasingly important as mechanisms for securing rights or enhancing access to collectively held lands. Organized social alliances, such as user group networks or Indigenous peoples' associations, and ad-hoc alliances among disparate social actors have also been instrumental in furthering reforms to enhance tenure security (see also Monterroso et al. 2017; Stoian et al. 2018; Gynch et al. 2020; Sharma et al. 2020).

### 2.5.1 RIGHTS RECOGNITION AND STRENGTHENING

In some areas, benefits have resulted from tenure reforms aimed at strengthening rights to collectively held resources. Larson and colleagues (2019) found that since the recognition of collective rights in Indonesia, Peru, and Uganda, more than three-quarters of community members felt their rights to land and forests were secure, and were confident their rights would be enforced. Community-owned forests and forests designated for community use by the state in these countries had similar outcomes for tenure security. However, results were mixed when communities that experienced tenure reforms were compared with control communities (that is, those whose lands were held under customary regimes that were not recognized statutorily). In Mozambique,

8 *Resguardos* are defined in Colombia's Decree 2001 of 1988 as "a special legal and sociopolitical institution composed of one or more indigenous communities that enjoys the guarantees of private ownership through collective titling, is in possession of its territory, and is governed for its management and internal affairs by an autonomous organization protected by indigenous law and its own norms" (cited in Arango 2018).

## TABLE 2 TYPES OF COLLECTIVELY HELD LANDS STUDIED BY COUNTRY

| Study site location | Type of collectively held land |
| --- | --- |
| **Latin America** | |
| Colombia | *Multiple types of collective tenure*, some with full recognition of ancestral rights and others that allow for collective titling without recognizing ancestral rights. The two dominant forms of collective titles are Indigenous resguardos, in which the Indigenous community that holds the title governs and manages its territory, and Afro-Colombian collective titles, in which Black communities are recognized as culturally distinct entities with standing similar to Indigenous peoples over their collectively held territories (Arrango 2018; Velasquez-Ruiz 2018). |
| Guatemala | *Community forest concessions on state land*. Local community associations can enter into 25-year contracts with the state to have the rights to manage and harvest forest products in protected areas (Gynch et al. 2020). Forest management plans must be approved by the state. |
| Mexico | *Indigenous community land and ejidos*. In the 1920s, Mexico recognized two categories of collectively held land: Indigenous community land and ejidos (Gynch et al. 2020). Indigenous community land consists of holdings where the customary rights of occupancy by Indigenous communities were formally recognized. Ejidos belong to non-Indigenous communities that have formally recognized rights to the land. Indigenous communities cannot sell their land, while communities that own ejidos can sell agricultural but not forest land (Lawry et al. 2012). Both types of tenure require state permits and approved management plans to harvest timber. |
| Peru | *Native community titles*. These are collective titles where the state formally recognizes the rights of Indigenous communities to land that has been delimited and registered with the state (Monterroso et al. 2017). These communities must also enter into a contract with the state to harvest timber and other commercial products in forests. The state retains lease rights to forests in areas covered by native community titles and can issue a concession for those lands to private enterprises (Monterroso et al. 2017). |
| **Africa** | |
| Ethiopia | *Community forests and participatory rangeland management*. Under Ethiopia's 2018 National Forest Law, communities can obtain titles for forests on their communal lands (McLain et al. 2019). Their rights are restricted to management and use rights; the state retains ownership rights. Communities must obtain state approval of their forest management plans. Participatory rangeland management (PRM) is a co-management approach that was modeled after Ethiopia's participatory forest management program (Flintan et al. 2019). PRM brings together multiple stakeholders, including state and customary authorities, to coproduce a rangeland management plan. Once the plan is developed, rangeland communities enter into a formal management agreement with the local government. |
| Kenya | *Community forests*. In Kenya, community forests are state-owned forests where the state has entered into co-management agreements with legally registered community forest associations formed in forest-adjacent communities (Tibalazika et al. 2019). |
| Mozambique | *Community land demarcation* (that is, community land titling). Mozambique's 1997 Land Law retained state ownership rights to land but provided the opportunity for communities to formalize their customary use rights to their land through georeferenced points outlining the boundaries of the area used by the community (Quan, Monteiro, and Mole 2013). The community cannot sell the land, but it can agree to let private individuals apply for a government leasehold to the area (Quan, Monteiro, and Mole 2013). |

| Study site location | Type of collectively held land |
|---|---|
| Namibia | *Wildlife* conservancies. Under Namibia's community-based natural resource management program, residents in communal areas can establish state-registered wildlife conservancies that are managed by a local conservancy management committee (Gynch et al. 2020). The communities have management and use rights to the conservancy land; many of the more successful conservancies are managed through joint venture agreements with private enterprises (Gynch et al. 2020). |
| Tanzania | *Certificates of customary occupancy rights*. Tanzania's 1999 Village Land Act authorizes villages to develop participatory village land use plans, including joint village land use plans (JVLUPs) encompassing lands used by multiple villages (Mwita, Kalenzi, and Flintan 2017). Once the plan is developed, the state issues certificates of customary occupancy rights, which grant use rights to the land user associations charged with managing the area in the JVLUP. |
| Uganda | *Community forests on state and customary lands, and collaborative forest management*. Uganda's 2003 National Forestry and Tree Planting Act authorized several approaches to ensuring the involvement of communities in forest management (Mukasa et al. 2019). One approach involves state-recognized community land associations entering into agreements with the National Forest Authority to manage nearby state forest reserves. For community forests established on customary land, customary forest management institutions hold management authority and customary rules apply. Collaborative forest management in the Ugandan context involves local user groups entering into a co-management agreement with the National Forest Authority. |
| **Asia** | |
| Indonesia | *Social forestry schemes*. Indonesia's 1999 Forestry Law No. 41 authorized two major collective types of social forestry schemes, in which communities or community groups could formalize various types of rights. These include *community forests and village forests*. The community forest program allows groups of farmers to enter into 35-year contracts with the national forestry service to manage production forests, and includes the right to harvest forest products. Village forests involve 35-year leases between village governance institutions and the national forestry service to protect and manage state-owned forests. |
| Nepal | *Community forests*. During the 1990s, Nepal enacted forest law reforms that enabled community forest user groups to obtain use rights to lands previously managed by the state forestry service (Sharma et al. 2020). The forest user groups are responsible for managing their forests and enforcing forest use rules (Lawry et al. 2012). They may sell forest products, but must develop a state-approved forest management plan (Sharma et al. 2020). |

more than 90 percent of communities included in a study of CLD outcomes experienced fewer land conflicts, both within and between communities, following the CLD process (Ghebru, Pitoro, and Woldeyohannes 2015). Perceptions of tenure security were much higher in CLD communities than in non-CLD communities.

Many studies showed that the intended beneficiaries of reforms are often unable to exercise their newly recognized or granted rights. The drivers of tenure insecurity identified earlier in this section—weak law enforcement, lack of awareness of rights, limited coordination across sectors, land conflicts, and social norms—also typically hindered the implementation of reforms. For collective tenure, additional barriers to reform implementation included (1) overly strict regulations together with complex bureaucratic procedures and (2) bureaucratic practices and politics.

**Overly strict regulations and complex bureaucratic procedures:** In identifying sources of tenure insecurity, all of the studies on collectively held land noted onerous regulations and time-consuming bureaucratic procedures for developing management or land-use plans and obtaining titles or certificates. In Peru, more than 38 steps are required for Indigenous communities to obtain title to their territories (Monterroso and Larson 2018a). In Nepal, regulations restricting commercial operations in community forests limit the ability of community enterprises to obtain enough raw material at a sufficiently low cost for wood product operations to be profitable (Sharma et al. 2020).

**Bureaucratic practices and politics:** Research on reform implementation revealed that land and forest administration infrastructure was inadequate to produce reforms, particularly at the local level. In Peru, regional land administration offices have neither the budget nor the personnel with the skills necessary to do their jobs. The lack of resources at the local government level is exacerbated by differences in regional and national priorities (Monterroso et al. 2019a). Tensions also exist between regional and national levels of governance over the responsibility for implementation, an issue that has been exacerbated as decentralization coincides with renewed efforts to implement native community titling (Monterroso et al. 2019a). Similarly, in Tanzania, village district councils lack the resources and staff to undertake VLUP (Mwita, Kalenzi, and Flintan 2017). In some cases, lack of clarity in government responsibilities was accompanied by a lack of transparency and accountability, creating governance conditions ripe for corruption (Dungumaro and Amos 2019; Larson, Monterroso, and Vigil 2019; Mwita, Kalenzi, and Flintan 2017).

## 2.5.2 MULTISTAKEHOLDER PROCESSES FOR TENURE SECURITY ENHANCEMENT

Multistakeholder processes (MSPs) are another mechanism with potential to enhance tenure security and expand access to resources, often through reducing conflicts over land or resources (Ratner et al. 2017a; 2017b; 2018). MSPs bring diverse stakeholders together to achieve particular goals that could not be achieved by the same stakeholders acting independently—whether it be managing conflict over resources, advocating for tenure reforms, ensuring that reforms are implemented as intended, or facilitating social learning (Ansell and Gash 2008). The objectives of most PIM-supported MSPs include enhancing tenure security and ensuring equitable land and resource access, such as through reducing conflicts (Mwita, Kalenzi, and Flintan 2017), identifying pathways to secure tenure (Zamora and Monterroso 2019), or expanding participation in land use planning and management (Flintan et al. 2019; Mwita, Kalenzi, and Flintan 2017). In all cases, the MSPs facilitated learning among participants at multiple scales (including individual, group, and network) with the objective of changing behavior and social relationships, a process known as social learning (Shelton, Janssen, and Meinzen-Dick 2018). This section describes two MSPs established for land use planning and two focused on identifying pathways for securing tenure and improving governance.

### 2.5.2.1 Participatory land planning and management

***Participatory rangeland management***

In Ethiopia, International Livestock Research Institute (ILRI) researchers affiliated with PIM5 assessed the impact of participatory rangeland management (PRM) on rangeland management and tenure security in two pastoral regions during 2017 and 2018 (Flintan et al. 2019). They examined two distinct forms of PRM: an early form and a later form, in which implementers had the opportunity to learn from and improve upon the earlier project. The initial PRM project used kebele boundaries as the rangeland management unit, and assisted communities in establishing associations with formal legal standing as the community-level governance units. The later PRM project used the full extent of each community's customary rangelands as the management unit, which covered a much larger area than one kebele. They worked with communities to revise and strengthen existing customary governance institutions rather than creating a new institution. In both projects, a multistakeholder pastoralist forum was established at the zonal level to address conflicts, provide input into rangeland governance decisions, and serve as a forum where pastoralists' concerns could be discussed with government representatives from multiple administrative levels.

Respondents reported that both projects improved livestock health, range conditions, livelihoods, and tenure security over rangelands. However, the long-term sustainability of PRM is not assured, because the process remains separate from government planning processes, and there is no legal framework specifically aimed at supporting PRM, thus making it unlikely that local governments will include it in their budgets (Flintan et al. 2019). The woreda participatory land-use planning, a land-use planning approach, is being piloted by Ethiopia's Ministry of Agriculture to address the need to embed rangeland management into governmental structures (Ethiopia, Ministry of Agriculture 2018).

***Joint village land use planning***

Although MSPs can be helpful for conflict management, the exclusion of key stakeholders can exacerbate existing conflicts or catalyze latent ones. This issue was illustrated by several examples from Tanzania. In an action-research project piloting the use of joint village land use planning (JVLUP), a state-recognized process in which villages decide how to manage shared commons, negotiations at one site collapsed due to conflicts between agriculturalists and pastoralists over the size of grazing areas (Mwita, Kalenzi, and Flintan 2017). At the other site, a joint grazing committee was formed to create a resource use plan, after which they applied for a certificate of customary rights occupancy. Although the decision-making was meant to be inclusive, in practice, it was dominated by village elites. Many villagers were unfamiliar with joint land use planning and became frustrated with the length of the process. Scaling JVLUP to other areas is in question, since neither the local nor the national government has budgeted for more plans, and replication depends on the availability of donor and NGO funding (Mwita, Kalenzi, and Flintan 2017).

### 2.5.2.2 Identifying pathways for securing tenure and improving governance

***Participatory prospective analysis***

As part of the Center for International Forestry Research's (CIFOR) Global Comparative Study on Forest Tenure Reforms, researchers piloted participatory prospective analysis (PPA), a multistakeholder futures scenario building and action plan development process in Colombia (Ortiz-Gerrero et al. 2018), Indonesia (Liswanti et al. 2019), Kenya (Mukasa et al. 2019b; Tibalazika et al. 2019), Nepal (Banjade, Paudel, and Mwangi 2020), Peru (Zamora and Monterroso 2019), and Uganda (Mshale et al. 2019a, 2019b). PPA consists of three stages, all of which take place in participatory workshops (Bourgeois et al. 2017). During the first stage, participants develop a list of the "forces of change" that result in tenure insecurity and then identify the driving forces of tenure insecurity. In the second stage, participants consider how the different driving forces might change over time and then develop a set of plausible scenarios of what the future might hold. In the third stage, participants analyze pathways to the possible scenarios, and then identify actions to forestall negative futures and enhance tenure security. The PPA workshops successfully engaged a diverse group of social actors in developing a list of driving forces, creating narratives for multiple possible futures, and identifying potential pathways to improve tenure security. Clear differences emerged among different actors (such as NGOs, government, and community leaders) over how tenure security can be achieved, illustrating the utility of PPA to facilitate a diversity of perspectives.

***Bayesian belief network analysis***

In Tunisia, Frija and colleagues (2019) applied Bayesian belief network analysis to model causality pathways for achieving good rangeland governance. Factor weights in the model are varied to compare how different scenarios affect the probability of achieving good governance for different land tenure categories (such as private, collectively owned, and public lands used collectivity by many tribes) and levels of previous infrastructure investment. The researchers organized workshops to develop the network structure, identify relevant governance indicators, and assign weights to the indicators. Through these workshops, data were collected from farmers as well as government representatives. Additional stakeholder workshops were held to validate and discuss the findings, a step that proved to be critical for finetuning the research questions and hypotheses. The two factors that most strongly influenced rangeland governance were the management performance of farmer associations and the relationship between the associations and land management councils, followed by tenure type and investment category. Frija and colleagues concluded that it is possible to improve rangeland governance for collective land if investments are made to expand the reach and organizational capacity of farmer associations and land management councils. Data from the workshops informed the design of Tunisia's pastoral code reforms (Werner et al. 2018). The model is part of a larger toolkit for rangeland restoration, which the International Union for the Conservation of Nature intends to pilot in Egypt and Jordan.

## 2.6 WOMEN'S LAND RIGHTS AND TENURE (IN) SECURITY: CROSS-CUTTING RESEARCH

PIM has supported an extensive series of studies on tenure security issues specific to women in multiple contexts and at various stages in their life cycle (see Billings, Meinzen-Dick, and Mueller 2014; Quisumbing and Kumar 2014; Pradhan, Meinzen-Dick, and Theis 2019). A detailed discussion of women's tenure security is beyond the scope of this synthesis. We focus here on the conceptual framework developed by Doss and Meinzen-Dick (2020) to facilitate the identification of entry points for strengthening women's land rights, and a real-world application of the framework to the study of tenure reforms in Indonesia, Peru, and Uganda.

Doss and Meinzen-Dick (2020) proposed a framework with four components: (1) the context in which women's land rights exist, (2) threats and opportunities to women's rights, (3) the "action arena," which consists of the social actors who affect women's tenure security, together with the resources available to those actors, and (4) the desired outcome, which is enhanced land tenure security for women. Collecting data on each component, and examining how the components interact, enables researchers and practitioners to more easily identify where and which types of tenure reforms or accompanying measures are most likely to strengthen women's land rights. The framework encourages researchers and practitioners to consider the range of factors and actors that can affect women's rights, enhancing the likelihood that factors requiring social change will be identified through the analysis, as well as those that can be addressed by legal and policy reforms.

Monterroso and colleagues (2019b) tested Doss and Meinzen-Dick's framework in a mixed methods comparative case study that examined how the outcomes of community forest reforms differed by gender in Indonesia, Peru, and Uganda. They looked at three types of reforms: (1) state lands designated for community use (Indonesia and Uganda), (2) state lands designated for company use (Indonesia), and (3) recognition of community ownership rights (Peru and Uganda). In all three countries, gender equity was inadequately incorporated into legal frameworks, women had limited roles in decision-making, and forest management committees did a poor job of incorporating women's concerns when developing forest use and management rules.

Monterroso and colleagues (2019b) concluded that achieving gender equity in community forest tenure reforms requires explicitly specifying that women are subjects of the reform. They emphasized the importance of incorporating language that provides explicit guidance on how to include women's concerns when implementing reforms. Moreover, they argued that involving women in decision-making forums is insufficient for gender parity—women also need to take an active role in conflict resolution, and extension services should be structured to meet women's needs. In many community forest reforms, monitoring and evaluation systems often do not measure impacts along gender lines, a weakness that suggests gender-differentiated evaluations should be made standard practice.

The study by Monterroso and colleagues showed that legal recognition of women's rights does not guarantee that women will be able to exercise those rights, a conclusion that is supported by other scholars (Quisumbing and Kumar 2014). It is therefore critical to identify and consider other entry points that affect how women participate and benefit from tenure reforms. Suggested remedies include programs that build the capacity of women to participate more extensively in formalization processes, as well as gender awareness training for government officials, NGOs, and communities (Monterroso et al. 2019b).

# 3. TOOLS FOR ASSESSING TENURE SECURITY AND REFORM IMPLEMENTATION

PIM-supported researchers collaborated with a variety of partner organizations to develop, refine, and/or adapt several tools for assessing tenure security, enhancing accountability for reform implementation, or both. The tools fall into four categories: conceptual frameworks; sets of indicators; training or educational tools, such as guides, manuals, and datasets; and analytical approaches.

## 3.1 CONCEPTUAL FRAMEWORKS

Selected examples of conceptual frameworks developed with PIM support are described in Table 3.

**TABLE 3 CONCEPTUAL FRAMEWORKS FOR ASSESSING TENURE SECURITY AND IDENTIFYING ENTRY POINTS FOR REFORMS**

| Framework | Description | Application |
|---|---|---|
| Framework for Monitoring and Evaluation of Land for Africa (Ghebru 2017) | Adaption of World Bank's Land Governance Assessment Framework for use in monitoring progress on land governance in Africa. Modified by adding indicators to measure progress on land rights for women, pastoralists, and other vulnerable groups. | Piloted in Madagascar, Malawi, and Tanzania |
| Framework for women's rights analysis (Doss and Meinzen-Dick 2020) | Describes the key elements that affect women's tenure security and how those elements interact. Enables the identification of points where it is feasible to expand or strengthen women's rights. Draws on the Institutional Analysis and Development (IAD) framework.[9] | Applied to PIM5-supported study of how women have been affected by community forest reforms in Indonesia, Peru, and Uganda (Monterroso et al. 2019b) |
| Tenure diagnostic for forest landscape restoration (FLR) assessments (McLain et al. 2018a) | Rights actualization model for identifying where tenure is likely to favor or inhibit adoption of FLR interventions, and what types of tenure and governance reforms are needed to support large-scale restoration. Draws on the IAD framework. | Informed research on tenure and FLR in Madagascar, which was supported by PIM5 and GIZ (the German Agency for International Cooperation). |
| 5Capitals-G (Stoian and Rodas 2018) | Gender-responsive adaptation of the 5Capitals toolkit, which assesses the poverty impacts of value chain development. Assesses gender differences in asset endowments of household and collective enterprises, and identifies constraints on women linked to cultural norms and values. | Piloted on forest concession enterprises and members in Guatemala by PIM5-supported researchers and partners |
| Framework for assessing the enabling environment for community forest enterprises (CFEs) (Sharma et al. 2020) | Diagnostic tool for assessing whether an enabling environment is present for CFEs and identifying intervention points for improving the likelihood of success. Draws on Baynes et al.'s (2015) framework of five success factors for community forestry. | Developed and piloted in Nepal by PIM5-supported researchers and partners |

9 The Institutional Analysis and Development framework is an analytical tool developed by Elinor Ostrom and her colleagues in 2005 for the purpose of analyzing the institutional processes by which individuals and collectivities make choices regarding decision-making or actions to take. More information about the details of the IAD framework are provided by McGinnis (2011).

| Framework | Description | Application |
|---|---|---|
| CFE investment readiness pathway (Gynch et al. 2020) | Conceptual framework linking community forest rights devolution to financial investments and environmental and social outcomes. | Developed but not yet applied |
| Learning measurement framework (Shelton, Janssen, and Meinzen-Dick 2018) | Tool for guiding efforts to measure whether learning takes place in participatory processes, as well as whether participation in such processes leads to behavioral change at individual, group, and societal levels. | Developed but not yet applied |

### BOX 1 MONITORING AND EVALUATION OF LAND IN AFRICA

The Monitoring and Evaluation of Land in Africa (MELA) project was a PIM-supported activity that developed new indicators with the intention of applying them broadly. The MELA framework was developed in response to the African Union's 2009 Declaration on Land Issues and Challenges in Africa, which called for member countries to implement systems for tracking progress in their land governance commitments. The framework was piloted in Madagascar, Malawi, and Tanzania. Project participants opted to use the World Bank's Land Governance Assessment Framework as a starting point and then incorporated 12 additional indicators tailored to the African context.

The custom indicators consist of quantitative measures that allow participating countries to measure progress toward ensuring that women, Indigenous peoples, pastoralists, and other vulnerable groups have rights to land in practice, as well as in law, and to assess the extent to which landholders perceive that they have tenure security. Working groups have been established in each country to further customize the generalized MELA framework, so that it captures information that participants deem appropriate for their country's context. Members of the working groups serve as "champions" for the framework, with the expectation that their involvement will increase the likelihood that the framework will be applied.

Most of these frameworks are designed to assess the sociopolitical systems in which tenure systems are embedded, help identify system elements that pose constraints to enhancing tenure security, and pinpoint where reform efforts should be targeted within the system. Section 1.6 and Box 1 provide examples of how these frameworks can be used.

## 3.2 INDICATORS

Indicators are useful tools for measuring progress toward a set of goals. In the Monitoring and Evaluation of Land in Africa (MELA) framework project (Box 1), PIM-supported researchers worked with national monitoring and evaluation teams in Africa to adapt the World Bank's Land Governance Assessment Framework indicators to different country contexts (Ghebru 2017). As with the MSP projects, the MELA project used a participatory approach to develop the indicators. Stakeholder participation in indicator development helped create a sense of ownership at the national level and ensured that indicators were locally relevant (Ghebru 2017).

## 3.3 TRAINING AND EDUCATIONAL TOOLS

A key pillar in the theory of change underlying PIM's work on tenure involves building capacity among social actors from multiple sectors and levels of governance to conduct research on the drivers and consequences of tenure security and the effectiveness of tenure reform implementation. To that end, PIM-supported researchers have developed an extensive set of training and educational materials to complement field research and engagement activities. These materials range from participatory land use and management planning manuals to step-by-step guides for Peruvian native communities seeking to acquire native community titles, to collective action games.

## 3.4 POLITICAL ECONOMY ANALYSIS: IDENTIFYING WINNERS AND LOSERS IN TENURE REFORM

The political economy of a country or region greatly influences how the costs and benefits of tenure reforms are distributed (Resnick 2021). Political economy analysis can help researchers and policymakers identify potential winners and losers of particular political choices and assess how those choices are likely to affect development interventions. Political economy analyses are thus useful for: (1) assessing the feasibility of policy reforms or institutional changes, (2) understanding why certain reforms are not implemented or have unintended consequences, and (3) identifying leverage points for policy change and the types of support needed to enable reform implementation (Resnick 2021).

Using political economy analysis, Suhardiman, Keovilignavong, and Kenney-Lazar (2019) found that within villages in highland areas of Lao PDR, tensions exist between families with long-standing claims to land and relative newcomers who must gain access to land through the first-occupant families. These tensions have led to gaps between national and village-level land-use planning objectives. The authors argue that these gaps persist because they benefit local community elites and government officials at multiple levels. At the same time, farmers and communities resist aspects of national land laws and land-use plans that inhibit their ability to farm upland areas, notably the prohibition on swidden agriculture,[10] which jeopardizes their food security.

Monterroso and colleagues (2017) used a political economy lens to trace the trajectory of reforms affecting land and forest tenure for Indigenous peoples in the Peruvian Amazon. By examining the political context underlying regulatory changes and their outcomes, the authors showed that several laws have significantly affected Indigenous tenure rights, whether intentionally or not. Rights recognition efforts began in the early 1970s and gained momentum during the 1980s and 1990s as social movements supporting Indigenous rights became more influential. However, it was not until the late 2000s, after several violent confrontations occurred between government representatives and Indigenous peoples, that broad public support emerged for implementing reforms that strengthened Indigenous collective rights, and progress was made toward native community titling.

The Lao PDR and Peru studies illustrate that tenure and governance reforms cannot be achieved solely through changes in law and policy. Analyses of how social actors relate to each other in terms of power and social standing can uncover how proposed reforms are likely to affect more powerful social actors and, if they are negatively affected, whether they are likely to resist or undermine reform implementation. Steps can then be taken to prevent resistance or bolster the reforms against attempts to undermine them.

10 Swidden agriculture, sometimes called shifting cultivation or, pejoratively, slash-and-burn farming, consists of a farming technique in which forested land is cleared, often through the use of fire, farmed for several years, and then left fallow so that the forest can regenerate.

# 4. KEY THEMES AND POLICY IMPLICATIONS

This section examines key themes articulated in PIM-supported research and their policy implications. The themes include: (1) the uneven distribution of tenure security and sources of tenure security for agricultural holdings, (2) impacts of tenure security on poverty, food security, and nutrition, (3) obstacles to tenure reform and implementation, and (4) the importance of social networks in effectuating and furthering tenure reforms.

## 4.1 UNEVEN DISTRIBUTION OF TENURE SECURITY AND SOURCES OF TENURE INSECURITY

Prominent factors associated with tenure insecurity in Ghana (Ghebru and Labrecht 2017), Mozambique (Ghebru and Girmachew 2019), Nigeria (Ghebru and Girmachew 2017), and Uganda (Doss, Meinzen-Dick, and Bomuhangi 2014) included urbanization, expanding land markets, population pressure, in-migration, and agricultural commodification. Across these countries, these factors tended to have disproportionately negative effects on tenure security for vulnerable groups, such as women and migrants. In Nigeria and Mozambique, sources of insecurity differed between men and women (Ghebru and Girmachew 2017; Ghebru and Girmachew 2019). Moreover, in both countries, perceptions of tenure insecurity and sources of insecurity differed depending on whether a woman was the head of household or a spouse in a male-headed household (Ghebru and Girmachew 2017; Ghebru and Girmachew 2019). These findings highlight the importance of understanding intra-community and intra-household dimensions of tenure when developing tenure security initiatives (Ghebru and Girmachew 2019). Across cases in SSA, tenure-secure households represented a substantial percentage in areas where customary authorities retain widespread legitimacy (Ghebru, Khan, and Lambrecht 2016; Lambrecht and Asare 2016; Ghebru and Girmachew 2017; Ghebru 2019). This finding suggests that in contexts where the state lacks land administration capacity, but customary authorities retain legitimacy, strengthening customary or hybrid systems might be a viable alternative to delivering tenure security.

## 4.2 OBSTACLES TO TENURE REFORMS AND REFORM IMPLEMENTATION

We highlight four obstacles to tenure reforms and reform implementation identified by PIM-supported research: (1) entrenched interests opposing reforms, (2) limited government capacity to implement reforms, (3) incompatibilities between social norms and reforms, and (4) resistant or corrupt bureaucratic cultures.

**Entrenched powerful interests:** Powerful rural and urban elites with entrenched interests in retaining the status quo pose an obstacle to tenure reforms. Political economy studies in Colombia (Ortiz-Guerrero et al. 2018), Lao PDR (Suhardiman, Keovilignavong, and Kenney-Lazar 2019), Myanmar (Suhardiman, Bright, and Palmano 2019), and Peru (Monterroso et al. 2017) indicate that economic development policies that favor converting forests to agriculture, plantations, or mineral extraction can be important counterforces to the implementation of reforms that favor collective tenure rights to land and forests. Attempts to harmonize conflicting sectoral policies are likely to encounter resistance from powerful elites, and strategies aimed at shifting the balance of power toward smallholders, forest dwellers, and pastoralists may be needed.

**Insufficient implementation capacity on the part of national and local governments:** Developing tenure reform policies and laws is one task, but implementing those reforms is entirely another undertaking. National and subnational governments lack the capacity to implement many of the tenure reforms described in this synthesis. This is true for reforms that affect collective tenure (Monterroso and Larson 2018a; Gynch et al. 2020; Sharma et al. 2020), as well as those aimed at increasing tenure security and access to individual/household land (Ghebru 2019). In a review of land governance frameworks in 10 countries in SSA, Ghebru (2019) found that only two countries had adequate resources to implement their land reform programs.

**Incompatibilities between social norms and reforms:** Comparative studies by Monterroso and colleagues (2019b) in Peru, Indonesia, and Uganda, and by Ghebru (2019) in Ethiopia, Malawi, Mozambique, and Nigeria, as well as single-country case studies in Tanzania (Dungumaro and Amos 2019) and Ethiopia (Quisumbing and Kumar 2014), found that tenure reform implementation was challenged by the presence of social norms that ran counter to newly enacted laws, regulations, and policies. State laws that provide protections for women's rights to land within marriage or consensual unions, and equitable inheritance rights for wives and female children are associated with more favorable outcomes for women's tenure security (see Kumar and Quisumbing 2015). However, customary norms that limit women's land rights still prevail in many areas, making it difficult for women to realize their rights under state law (Dungumaro and Amos 2017; Ghebru 2019; Lambrecht and Asare 2017; Quisumbing et al. 2014).

**Resistant or corrupt bureaucratic cultures:** Resistance from government officials charged with implementing reforms was another obstacle to reform implementations that cut across studies of tenure on agricultural holdings as well as collective tenure (Dungumaro et al. 2017; Ghebru 2019; Monterroso and Larson 2018a; Suhardiman, Bright, and Palmano 2019). In some cases, corruption within government bureaucracies further undermined reform implementation (Ghebru 2019; Monterroso and Larson 2018a; Suhardiman, Bright, and Palmano 2019).

## 4.3 A KEY ROLE FOR SOCIAL NETWORKS

A key takeaway from PIM-supported tenure research is the critical importance of engaging a broad array of social actors and negotiating the power imbalances that inevitably exist when developing and implementing reforms to provide tenure security for rural inhabitants and communities (Gynch et al. 2020; Monterroso et al. 2017; Ratner, Rivera, and Fiorenza 2019; Suhardiman, Bright, and Palmano 2019; 2019b). Tools such as political economy analysis can be deployed to understand the power dynamics within the action space, identify the motivations of different social actors, and determine who can influence which elements of the tenure and governance system in ways that support desired reforms (Suhardiman, Bright, and Palmano 2019; Suhardiman, Keovilignavong, and Kenney-Lazar 2019). Broad-based MSPs are promising tools to help move from enacting reforms to ensuring their implementation; these are oriented specifically toward identifying and influencing those levers for change (Ratner, Rivera, and Fiorenza 2019). Eight years of PIM-supported research on tenure security and related governance issues has helped build a deeper understanding of the roles that different social actors play in advancing—and in some cases, opposing or resisting—tenure reforms that expand and secure access to land and resources for rural populations. Table 4 lists some of the roles that different social actors play in strengthening tenure security.

## 4.4 TENURE (IN)SECURITY IMPACTS ON POVERTY, FOOD SECURITY, AND NUTRITION

Few of the PIM-supported studies explicitly addressed the impacts of tenure (in)security on poverty, food security, and nutrition. However, studies that explored these relationships suggest that greater tenure security is associated with lower levels of poverty and improved food security and nutrition. These relationships held true for agricultural holdings, including women's holdings,

**TABLE 4 SOCIAL ACTORS AND THEIR ROLES**

| Social actors | Roles |
|---|---|
| Community leaders and members | Advocate for (or against, in some cases) policy reforms; likely to be more effective when communities join forces to establish user group networks, such as the Federation of Community Forest Users in Nepal (FECOFUN) and Guatemala's Community Forest Association of Péten, and social movements, such as those initiated by Indigenous peoples in Colombia and Peru. Communities also carry out functions related to land and resource allocation, management, and conflict resolution. In some cases, they may oppose land reforms or resist implementation if the reforms undermine their political or economic power. |
| Governments (multiple scales) | Establish new policies and laws that (potentially) strengthen tenure security and reform implementation, backstopping communities as they seek to enforce their rules; offer service delivery, such as extension and technical assistance; and provide conflict management and resolution when local systems cannot resolve conflicts. In some cases, they may oppose land reforms that limit their power or interfere with their ability to profit from rent-seeking in dysfunctional systems. |
| Donor agencies | Provide significant funding and technical assistance to governments, NGOs, research institutions, and contractors; place pressure on governments to implement reforms (though these efforts are misguided, in some cases); and convene national, regional, and international dialogue forums to advance tenure security. |
| Nongovernmental organizations (NGOs) (operating at multiple scales) | Typically function as brokers between national and subnational governments and rural inhabitants; often act as advocates for reforms (or against, in some cases); invest in social network expansion by linking groups together; and provide funding and various services to communities, including extension, conflict resolution and management, and legal aid. |
| Research institutions | Identify gaps in knowledge about tenure systems, develop conceptual and practical tools to advance knowledge about tenure security; collect and analyze data; develop training, educational materials, and other knowledge products; provide training and opportunities for learning; convene knowledge-sharing forums; and may function as brokers and network builders linking various social actors. |
| Private sector firms | Provide financing for community forestry in some countries; may advocate for reforms (or against, in some cases); provide support for market-based approaches to strengthening rights (such as certification processes for forest products); and develop and implement voluntary guidelines for respecting land rights in supply chains. In some cases, they may oppose land reforms if the changes undermine their economic or political power. |

collectively managed lands, and across multiple livelihood sectors linked to natural resources.

**Agricultural holdings:** On agricultural holdings in Ethiopia, secure tenure was associated with higher agricultural productivity (Ghebru and Holden 2015), lower levels of poverty (Ghebru and Holden 2019), and more investment in land conservation measures (Quisumbing and Kumar 2014). More secure tenure for agricultural holdings was also associated with better nutrition for young children in the Kyrgyz Republic (Kosec and Shemyakina 2019). However, the mechanisms that link tenure security with food security and improved nutrition require greater specification.

**Women's tenure security:** Reviewing the evidence on how women's land tenure security can contribute to poverty reduction, Meinzen-Dick and colleagues (2019) found a positive link between women's land rights and empowerment, natural resource management, and food security. However, more evidence is required to identify the contributions of women's land rights to credit access, technology adoption, and agricultural productivity.

**Collectively managed land:** Implementation of PRM in Ethiopia resulted in better ecological conditions, which in turn were associated with improvements in livestock conditions and health of community members (Flintan et al. 2019). Likewise, Hellin and colleagues (2018) reported that food security improved among farmers in Guatemala after a protected area was established, over which the local community had co-management rights. However, Larson and colleagues (2019) found no statistically significant difference for food security between communities with collective forests and those in control communities in Indonesia, Peru, and Uganda.

**Fisheries:** In a study encompassing protected marine areas in the Atlantic, Indian, and Pacific oceans, McLain and colleagues (2018b) found that community-managed reserves had better outcomes for fish populations—presumably with a positive impact on fishers' livelihoods—than protected areas managed by national governments, largely because the community-managed reserves had more effective rule enforcement (McLain et al. 2018b). MSPs that reduced conflicts and improved resource management also yielded improved conditions for fisheries in Cambodia, Uganda, and Zambia (see Ratner et al. 2017a; 2018). These studies suggest that tenure security can lead to improved ecological outcomes. However, an explicit link remains missing between ecological outcomes and improvements in food security and nutrition.

## 4.5 POLICY IMPLICATIONS

The PIM-supported studies yielded a wealth of knowledge about tenure security for agricultural holdings and resource commons, including whose tenure is (in)secure, the sources of tenure insecurity and how they differ within communities and households, the types of mechanisms being used to strengthen tenure security, and challenges with implementation. From this work, we highlight the following policy implications for tenure reform efforts:

**Tailor tenure interventions to a range of contexts.** A key message from the research on agricultural holdings is the need to move away from a one-size-fits-all approach to interventions that enhance tenure security (Ghebru, Khan, and Lambrecht 2016; Ghebru and Girmachew 2017; Ghebru 2019; Lambrecht and Asare 2016). Instead, it is important that policymakers and designers of tenure security initiatives understand differences in tenure security, both geographically and among population subgroups within their sphere of operations. This knowledge can enable the design of interventions and accompanying measures to fit the tenure circumstances of different regions, and enhance security for a broad range of landholders and land users, including subgroups for whom tenure insecurity is most acute.

**Where feasible, adopt "bottom-up" tenure reforms.** Greater emphasis needs to be placed on developing "bottom-up" approaches that focus on reform from within customary or hybrid tenure and governance systems, rather than assuming that state-run land registration systems are the only way to achieve tenure security (Ghebru, Khan, and Lambrecht 2016; Ghebru 2019; Ranjatson et al. 2019). Bottom-up tenure reforms may be a good alternative in remote areas where demand for land is relatively low, customary governance systems remain functional, and the state lacks the resources and technical skills to implement and maintain an accessible land registration system. However, because tenure security is typically much weaker for some groups, notably women, youth, migrants, and pastoralists, mechanisms for improving security for marginalized groups should be integrated into such approaches. These mechanisms may include capacity building to enable vulnerable groups to participate in the design and implementation of tenure reform, legal aid and empowerment programs, and broad-based initiatives to raise legal awareness.

**Incorporate accompanying measures that enable women to realize their rights.** In areas where customary norms limit women's tenure rights, additional measures need to accompany tenure reforms, such as capacity building to enable women to participate in the design and implementation of tenure reform, legal rights awareness campaigns, and women's empowerment programs. Measures that include men in awareness-raising efforts are equally critical, as well as those that provide training in women's rights for judges and customary authorities responsible for resolving land disputes.

**Support social movements, user group networks, and MSPs to address power imbalances.** Sustained and extensive support for social movements and user group networks has proved successful in shifting power balances to a point where reforms are possible (Monterroso et al. 2017; Gynch et al. 2020). MSPs modeled after the International Land Coalition's National Engagement Strategy, which takes a broad-based approach to influencing reform processes, have also been successful (Ratner, Rivera, and Fiorenza 2019). However, both strategies require long-term investments in building and sustaining social networks at multiple scales (Gynch et al. 2020; Monterroso et al. 2017; Ratner, Rivera, and Fiorenza 2019).

**Use tenure security enhancement as a tool to reduce poverty, and improve food security and nutrition.** Although limited, the evidence suggests that tenure enhancement reforms, whether aimed at agricultural holdings or collectively managed land and resources, can play an important role in poverty reduction and enhanced food security and nutrition goals. However, additional research is needed to better understand how tenure security affects food security and nutrition.

# 5. CONCLUSION: LESSONS LEARNED FOR SUPPORTING REFORMS AND RESEARCH PRIORITIES

Research supported by PIM's Flagship 5 Program has contributed substantively to the body of scientific knowledge on tenure and governance.

Importantly, this research has helped advance policy reforms that improve access and tenure security of land and natural resources, promoted more inclusive and equitable decision-making, and supported more equitable and sustainable use of resources in Africa, Asia, and Latin America. It has accomplished this by empowering and strengthening the capacity of a variety of actors, including government, donors, civil society organizations, NGOs, community authorities, and community members, to engage in tenure and governance research and implementation.

The body of knowledge on tenure systems—particularly regarding when and how tenure rights serve as incentives (or disincentives) to invest in conservation practices—developed through PIM-supported applied research and engagement is invaluable in contexts where conservation practices must be scaled over large areas. This will be particularly important for initiatives such as forest landscape restoration (Table 3), which requires collaboration between different actors and sectors operating at multiple governance and ecological scales.

We conclude with a discussion of key lessons learned from PIM's tenure work about how to engage effectively in policy change-oriented action research, followed by recommendations for research priorities as IFPRI advances its efforts to provide policy-relevant research in support of a world free of hunger and malnutrition.

## 5.1 KEY LESSONS FOR SUPPORTING TENURE REFORM AND IMPLEMENTATION

**Long-term partnerships are key to uptake.** PIM's achievements with respect to enhancing tenure security have come about through long-term, sustained partnerships with national and subnational governments, civil society, NGOs, and donor agencies. Implementing reforms that enhance tenure security requires building awareness of the importance of tenure and governance, and expanding the number and types of social actors engaged in tenure reform and implementation. It also requires investments in building trust among the expanding network of partners, and building the capacity of partners so that they can better adopt or adapt the tools developed through applied research.

**Ongoing engagement and training for actors at multiple scales increases the likelihood of successful reform implementation.** The importance of ongoing engagement and provision of training to a diverse set of social actors operating at multiple scales cannot be overstated. Engaging a broad set of social actors will bring diverse perspectives, values, skills, and knowledge to the table, and enhance the likelihood of buy-in at a scale that is sufficiently broad as to increase the likelihood of long-term success (Ratner, Rivera, and Fiorenza 2019). Training materials and programs in support of reform implementation are likely to be more effective if they are co-produced by researchers working in collaboration with the social actors meant to benefit from them. Box 2, which draws from PIM-supported work on tenure security in Peru, describes a co-produced training program on tenure, gender, and Indigenous issues.

**BOX 2 CHANGES TO UNIVERSITY-LEVEL FORESTRY CURRICULUM IN PERU**

PIM5 and the forestry department of the National Agrarian University in Peru partnered on a study of land and forest tenure security of Indigenous communities. The outcomes included increased knowledge and greater research capacity of university staff on tenure, gender, and Indigenous issues. In addition, the forest engineer curriculum was changed to include a mandatory social science concentration. In collaboration with CIFOR, the forestry department organized a three-day international congress in Lima in June 2018. A lead university researcher reported that participating in the event led her to change her teaching and add a new course on ethnoecology. Integrating such training into university-level forestry curriculum will have positive long-term consequences for improved land management, as it fills a knowledge and skills gap that PIM5 researchers have documented as posing a barrier to the implementation of native titling reform. More than 500 community members also participated in trainings on tenure rights, and each of the 22 villages participating in the study received a book summarizing the findings from their community, thus helping to fill a gap in community-level awareness of tenure rights.

**Strengthening private sector engagement in action research on tenure reforms needs more attention.** Infusing PIM5 research findings among private sector actors has proved challenging, perhaps because researchers' connections with the private sector, particularly large firms or industry associations, are much less strong than with other actors. Given the strong role that private sector actors play in the politics and economics of land and natural resource governance, it is worth investigating approaches for engaging these actors in ways that will enhance, rather than undermine, less powerful actors' tenure security and participation in governance. Additionally, engagement of the private sector will likely become increasingly important for reform implementation as community forest and rangeland enterprises mature, and their need for private sector investment expands (Gynch et al. 2020).

## 5.2 RESEARCH PRIORITIES

Based on the findings of the tenure studies supported through PIM between 2013 and 2021, we suggest four priorities for tenure security research aimed at informing efforts to achieve SDGs related to natural resource management, food security and nutrition, and poverty alleviation.

**(1) Investigate when, where, and for whom customary tenure systems provide security for lands held by individuals and households.** Identify and test mechanisms for enhancing the ability of these systems to provide tenure security in ways that are inclusive and equitable. This may require identifying culturally sensitive ways to change social norms, as well as harmonizing key elements of state and customary law. Areas needing particular attention include (1) the identification of best practices for ensuring that women are actively engaged in design, implementation, and monitoring (Monterroso et al. 2019b) and (2) the investigation of mechanisms for strengthening—rather than replacing—customary tenure systems in areas where customary institutions retain widespread legitimacy (Ghebru 2019; Ranjatson et al. 2019).

**(2) Explore how the structure, composition, and processes of multistakeholder platforms influence tenure security, livelihood, and ecological outcomes.** Multistakeholder platforms have emerged as promising tools for resolving conflicts and engaging in collective action for large-scale landscape restoration and conservation initiatives. However, the impacts of the different approaches employed by multistakeholder platforms on land governance, including tenure security, are not well-researched, and warrant further investigation (Ratner et al. 2018). It will be important to test frameworks used to map power structures and the social relationships that maintain them to see which are effective, and in what contexts, for identifying points of leverage and strategies that can lead to sustainable reforms. Finding mechanisms that can enable MSPs to successfully navigate extreme power imbalances and incorporate diverse types of knowledge into decision-making processes are other important areas of investigation.

**(3) Clarify how tenure security is linked to food security and nutrition.** The links between tenure security and poverty have been a focus of research for many decades (Oyono 2021), but the relationship of tenure security to food security (Holden and Ghebru 2016) and nutrition (Ibrahim, Hendriks, and Schönfeldt 2022) has been less studied. Additional research that fills these gaps is critical, as are studies that examine the nexus between tenure security, poverty, food security, and nutrition.

**(4) Invest in transformational socioecological research and engagement.** Meeting the SDGs will require implementing restoration projects on a very large scale, as well as engaging multiple actors, sectors, and disciplines. This calls for a long-term investment in research and engagement that emphasizes transdisciplinary approaches. PIM5-supported work during 2013–2021 provides several promising examples of what these investments might involve. Model projects include the PRM work in Ethiopia (Flintan et al. 2019), Tanzania (Mwita, Kalenzi, and Flintan 2017), and Tunisia (Frija et al. 2019), and the Promise of the Commons initiative in India (Box 3). Over the long term, these projects have the potential to produce positive ecological outcomes and improvements in livelihoods, but research is needed that documents whether and in what circumstances these models achieve their outcomes.

### BOX 3 PROMISE OF THE COMMONS INITIATIVE IN INDIA

PIM5 has built on long-term collaboration between IFPRI and the Foundation for Ecological Security in India to strengthen tenure and governance of the commons. This work has been expanded under the Promise of the Commons initiative, which seeks to identify ways to expand the area of common property restored or enhanced in India from 5 million to 20 million acres over five years. The Promise of the Commons includes more than 80 partner organizations in India, as well as IFPRI, CIFOR, World Agroforestry Centre, Bioversity International, and ILRI. In addition to helping communities secure rights to the commons, the initiative draws on PIM-supported guidance on inclusive facilitation processes, multistakeholder platforms, and use of resource management games as tools for experiential learning, which were developed by researchers funded through PIM.

# REFERENCES


Ansell, C., and Gash, A. 2008. "Collaborative Governance in Theory and Practice." *Journal of Public Administration Research and Theory* 18 (4): 543–571. https://doi.org/10.1093/jopart/mum032

Arango, J.H. 2018. *Collective Tenure in Colombia: Data and Trends*. CIFOR Infobrief 213. Bogor, Indonesia: Center for International Forestry Research (CIFOR). https://doi.org/10.17528/cifor/006877

Banjade, M.R., N.S. Paudel, and E. Mwangi. 2020. *Insights from Participatory Prospective Analysis (PPA) Workshops in Nepal*. CIFOR Infobrief 276. Bogor, Indonesia: CIFOR. https://doi.org/10.17528/cifor/007553

Baynes, J., J. Herbohn, C. Smith, R. Fisher, and D. Bray. 2015. "Key Factors which Influence the Success of Community Forestry in Developing Countries." *Global Environmental Change* 35: 226–238. https://doi.org/10.1016/j.gloenvcha.2015.09.011

Behrman, J., L. Billings, and A. Peterman. 2013. "Evaluation of Grassroots Community-based Legal Aid Activities in Uganda and Tanzania: Strengthening Women's Legal Knowledge and Land Rights." CAPRi Working Paper 108. International Food Policy Research Institute (IFPRI), Washington, DC. http://dx.doi.org/10.2499/CAPRiWP108.

Billings, L., R.S. Meinzen-Dick, and V. Mueller. 2014. *Implications of Community-based Legal Aid Regulation on Women's Land Rights*. IFPRI Research Brief 20. Washington, DC: IFPRI. http://ebrary.ifpri.org/cdm/ref/collection/p15738coll2/id/128173

Boone, C. 2019. "Legal Empowerment of the Poor through Property Rights Reform: Tensions and Trade-offs of Land Registration and Titling in Sub-Saharan Africa." *Journal of Development Studies* 55 (3): 384–400. https://doi.org/10.1080/00220388.2018.1451633

Boué, C., and J-P Colin,. 2018. "Land Certification as a Substitute or Complement to Local Procedures? Securing Rural Land Transactions in the Malagasy Highlands." *Land Use Policy* 72: 192–200. https://doi.org/10.1016/j.landusepol.2017.12.049

Bourgeois, R., N. Liswanti, C. Mukasa, A. Zamora, T. Herawati, I. Monterroso, B. Mshale, M.R. Banjade, E. Mwangi, and A. Larson. 2017. *Guide for Co-Elaboration of Scenarios: Building Shared Understanding and Joint Action for Reform and Security of Forest Tenure*. Bogor, Indonesia: CIFOR. https://doi.org/10.17528/cifor/006749

Chimhowu, A. 2019. "The 'New' African Customary Land Tenure. Characteristic, Features and Policy Implications of a New Paradigm." *Land Use Policy* 81: 897–903. https://doi.org/10.1016/j.landusepol.2018.04.014

Deininger, K., D.A. Ali, S. Holden, and J. Zevenbergen. 2008. "Rural Land Certification in Ethiopia: Process, Initial Impact, and Implications for Other African Countries." *World Development* 36 (10): 1786–1812. https://doi.org/10.1016/j.worlddev.2007.09.012

Doss, C.R., R. Meinzen-Dick, and A. Bomuhangi. 2014. "Who Owns the Land? Perspectives from Rural Ugandans and Implications for Large-scale Land Acquisitions." *Feminist Economics* 20 (1): 76–100. 10.1080/13545701.2013.855320

Doss, C., and R. Meinzen-Dick. 2020. "Women's Land Tenure Security: A Conceptual Framework." *Land Use Policy* 99: 105080. https://doi.org/10.1016/j.landusepol.2020.105080

Dungumaro, J., and M. Amos. 2019. *Pastoral Women's Rights and Leadership Forums, Tanzania: Experience, Impact and Lessons Learned*. Rangelands Research Report 3. Nairobi, Kenya: International Livestock Research Institute (ILRI). https://cgspace.cgiar.org/handle/10568/106430

Ethiopia, Ministry of Agriculture. 2018. *Woreda Participatory Land Use Planning (WPLUP) in Pastoral and Agro-pastoral Areas*. Volume I: Manual. Addis Ababa, Ethiopia: Government of Ethiopia. https://hdl.handle.net/10568/99262

Flintan, F., A. Ebro, B. Eba, A. Assefa, Y. Getahun, K. Reytar, B. Irwin, et al. 2019. *Review of Participatory Rangeland Management (PRM) Process and Implementation*. Rangelands Research Report 2. Nairobi, Kenya: ILRI. https://cgspace.cgiar.org/handle/10568/106017

Frija, A., M. Sghaier, B. Dhehibi, and M. Fetoui. 2019. *Frameworks, Tools, and Approaches for the Assessment of Rangeland Governance*. Tunis, Tunisia: International Center for Agricultural Research in the Dry Areas (ICARDA). https://hdl.handle.net/20.500.11766/10081

Gebregziabher, G., and S.T. Holden. 2011. "Distress Rentals and the Land Rental Market as a Safety Net: Contract Choice Evidence from Tigray, Ethiopia." *Agricultural Economics* 42 (supplement): 45–60. https://doi.org/10.1111/j.1574-0862.2011.00551.x

Ghebru, H. 2017. "Monitoring and Evaluation of Land in Africa (MELA) Project." Presentation, Malawi National Land Symposium, Lilongwe, Malawi, May 10. https://hdl.handle.net/10568/121585

Ghebru, H. 2019. "Women's Land Rights in Africa." In *Regional Strategic Analysis and Knowledge Support System (ReSAKSS) 2019 Annual Trends and Outlook Report: Gender Equality in Rural Africa: From Commitments to Outcomes*, eds. A. Quisumbing, R. Meinzen-Dick, and J. Njuki, 44–56. Washington, DC: IFPRI. https://doi.org/10.2499/9780896293649_04

Ghebru, H., and F. Girmachew. 2017. "Scrutinizing the Status Quo: Rural Transformation and Land Tenure Security in Nigeria." NSSP Working Paper 43. IFPRI, Washington, DC. http://ebrary.ifpri.org/cdm/ref/collection/p15738coll2/id/131363

Ghebru, H., and F. Girmachew. 2019. "Perceived Tenure (In)Security in the Era of Rural Transformation: Gender-disaggregated Analysis from Mozambique." IFPRI Discussion Paper 1799. Washington, DC: IFPRI. https://doi.org/10.2499/p15738coll2.133062

Ghebru, H., and F. Girmachew. 2020. "Direct and Spillover Effects of Land Rights Formalization in Africa: A Case Study of the Second-Level Land Certification (SLLC) in Ethiopia." IFPRI Discussion Paper 01968. Washington, DC: IFPRI. https://doi.org/10.2499/p15738coll2.134083

Ghebru, H., and S.T. Holden. 2015. "Technical Efficiency and Productivity Differential Effects of Land Right Certification: A Quasi-Experimental Evidence." *Quarterly Journal of International Agriculture* 54 (1): 1–31. https://econpapers.repec.org/RePEc:ags:qjiage:206294

Ghebru, H., and S. Holden. 2019. "Land Rental Markets and Rural Poverty Dynamics in Northern Ethiopia: Panel Data Evidence using Survival Models." *Review of Development Economics* 23 (1): 131–154. https://doi.org/10.1111/rode.12548

Ghebru, H., and A. Kennedy. 2019. *Nigeria Land Governance Reform: What Needs to be Done to Stimulate Demand and Support Market Growth?* Feed the Future Innovation Lab for Food Security Policy Research Brief 97. East Lansing, MI: Michigan State University. http://ebrary.ifpri.org/cdm/ref/collection/p15738coll2/id/133526

Ghebru, H., H. Khan, and I. Lambrecht. 2016. "Perceived Land Tenure Security and Rural Transformation: Empirical Evidence from Ghana." IFPRI Discussion Paper 1545. Washington, DC: IFPRI. http://ebrary.ifpri.org/cdm/ref/collection/p15738coll2/id/130495

Ghebru, H., and I. Lambrecht. 2017. "Drivers of Perceived Land Tenure (In)Security: Empirical Evidence from Ghana." *Land Use Policy* 66: 293–303. https://doi.org/10.1016/j.landusepol.2017.04.042

Ghebru, H., R. Pitoro, and S. Woldeyohannes. 2015. "Customary Tenure and Innovative Measures of Safeguarding Land Rights in Africa: The Community Land Initiative (Iniciativa de Terres Comunitárias) in Mozambique." IFPRI Discussion Paper 484. Washington, DC: IFPRI. http://ebrary.ifpri.org/cdm/ref/collection/p15738coll2/id/129826

Gynch, S., S. Lawry, R. McLain, I. Monterroso, and A. Adhikary. 2020. "Common Benefits: Is Community Tenure Facilitating Investment in the Commons for Inclusive and Sustainable Development?" *Forest Policy and Economics* 111: 102088. https://doi.org/10.1016/j.forpol.2019.102088

Hellin., J., B.D. Ratner, R. Meinzen-Dick, and S. Lopez-Ridaura. 2018. "Increasing Social-Ecological Resilience within Small-scale Agriculture in Conflict-affected Guatemala." *Ecology and Society* 23 (3): 5. https://doi.org/10.5751/ES-10250-230305

Holden, S.T., and H. Ghebru. 2016. "Land Tenure Reforms, Tenure Security and Food Security in Poor Agrarian Economies: Causal Linkages and Research Gaps." *Global Food Security* 10: 21–28. http://dx.doi.org/10.1016/j.gfs.2016.07.002

Ibrahim, K.H., S.L. Hendriks, and H. Schönfeldt. 2022. "The Effect of Smallholder Land Tenure on Child Malnutrition in Nigeria." *Land Use Policy* 119: 106214. https://doi.org/10.1016/j.landusepol.2022.106214

Kisambu, N., F. Flintan, E. Daley, and S. Pallas. 2017. "Gender and the Commons: Pastoral Women's Land Rights and Village Land Use Planning in Tanzania: Experiences from the Sustainable Rangeland Management Project." Presentation, Conference of the International Association for the Study of the Commons, Utrecht, the Netherlands, July 10–14. https://orcid.org/0000-0002-9732-097X

Kosec, K., and O.N. Shemyakina. 2019. "Land Reform and Child Health in the Kyrgyz Republic." IFPRI Discussion Paper 1904. Washington, DC: IFPRI. https://doi.org/10.2499/p15738coll2.133567

Kumar, N., and A.R. Quisumbing. 2015. "Policy Reform toward Gender Equality in Ethiopia: Little by Little the Egg Begins to Walk." *World Development* 67: 406–423. https://doi.org/10.1016/j.worlddev.2014.10.029

Lambrecht, I., and S. Asare. 2016. "The Complexity of Local Tenure Systems: A Smallholders' Perspective on Tenure in Ghana." *Land Use Policy* 58: 251–263. https://doi.org/10.1016/j.landusepol.2016.07.029

Larson, A.M., I. Monterroso, N. Liswanti, T. Herawati, A. Banana, P. Cantuarias, K. Rivera, and E. Mwangi. 2019. *Models for Formalizing Customary and Community Forest Lands: The Need to Integrate Livelihoods into Rights and Forest Conservation Goals.* CIFOR Infobrief 253. Bogor, Indonesia: CIFOR. https://doi.org/10.17528/cifor/007273

Larson, A.M., I. Monterroso, and N.H. Vigil. 2019. *Conflict in Collective Forest Tenure: Lessons for Peru from a Comparative Study*. CIFOR Infobrief 243. Bogor, Indonesia: CIFOR. https://doi.org/10.17528/cifor/007148

Larson, A.M., and J. Springer. 2016. "Recognition and Respect for Tenure Rights." NRGF Conceptual Paper. International Union for Conservation of Nature and Natural Resources (IUCN), Gland, Switzerland. https://hdl.handle.net/10568/93949

Lawry, S., R. McLain, B. Swallow, and K. Biedenweg. 2012. *Devolution of Forest Rights and Sustainable Forest Management. Volume 1: A Review of Policies and Programs in 16 Developing Countries*. Washington, DC: United States Agency for International Development. https://www.land-links.org/research-publication/devolution-of-forest-rights-and-sustainable-forest-management-volume-1-a-review-of-policies-and-programs-in-16-developing-countries/

Liswanti, N., E. Mwangi, M.R. Banjade, and T. Herawati. 2019. *What Future Direction for Forest Tenure Reform Implementation in Indonesia? Perspectives of National-level Stakeholders*. CIFOR Infobrief 256. Bogor, Indonesia: CIFOR. https://doi.org/10.17528/cifor/007283

McGinnis, M.D. 2011. "An Introduction to IAD and the Language of the Ostrom Workshop: A Simple Guide to a Complex Framework." *Policy Studies Journal* 39: 169–183. https://doi.org/10.1111/j.1541-0072.2010.00401.x

McLain, R., H. Kasse, S. Lawry, and B. Yazew. 2019. *Fostering Tenure Security for Forest Landscape Restoration in Ethiopia: Creating Enabling Conditions for the 2018 Forest Proclamation*. CIFOR Infobrief 267. Bogor, Indonesia: CIFOR. https://doi.org/10.17528/cifor/007410

McLain, R., S. Lawry, M.R. Guariguata, and J. Reed. 2018a. "Toward a Tenure-Responsive Approach to Forest Landscape Restoration: A Proposed Tenure Diagnostic for Assessing Restoration Opportunities." *Land Use Policy* 104: 103748. https://doi.org/10.1016/j.landusepol.2018.11.053

McLain, R., S. Lawry, and M. Ojanen. 2018b. "Fisheries' Property Regimes and Environmental Outcomes: A Realist Synthesis Review." *World Development* 102: 213–227. https://doi.org/10.1016/j.worlddev.2017.09.016

Meinzen-Dick, R.S., A.R. Quisumbing, C.R. Doss, and S. Theis. 2019. "Women's Land Rights as a Pathway to Poverty Reduction: A Framework and Review of Available Evidence." *Agricultural Systems* 172: 72–82. https://dx.doi.org/10.1016/j.agsy.2017.10.009

Monterroso, I., P. Cronkleton, D. Pinedo, and A.M. Larson. 2017. "Reclaiming Collective Rights: Land and Forest Tenure Reforms in Peru (1960–2016)." CIFOR Working Paper 224. CIFOR, Bogor, Indonesia. http://www.cifor.org/library/6426/reclaiming-collective-rights.desc-land-and-forest-tenure-reforms-in -peru-1960-2016/

Monterroso, I., and A. Larson. 2018a. *Challenges in Formalizing the Rights of Native Communities in Peru*. CIFOR Infobrief 231. Bogor, Indonesia: CIFOR. https://doi.org/10.17528/cifor/006294

Monterroso, I., and A.M. Larson. 2018b. *Progress in Formalizing "Native Community" Rights in the Peruvian Amazon (2014–2018)*. CIFOR Infobrief 233. Bogor, Indonesia: CIFOR. https://doi.org/10.17528/cifor/007080

Monterroso, I., A.M. Larson, J. Quaedvlieg, F. Valencia, L. Jarama, and J. Saldaña. 2019a. *Formalization of the Collective Rights of Native Communities in Peru: The Perspective of Implementing Officials*. CIFOR Infobrief 247. Bogor, Indonesia: CIFOR. https://doi.org/10.17528/cifor/007271

Monterroso, I., A. Larson, E. Mwangi, N. Liswanti, and Z. Cruz-Burga. 2019b. *Mobilizing Change for Women within Collective Tenure Regimes*. Bogor, Indonesia: CIFOR. https://www.cifor.org/knowledge/publication/7440/

Mshale, B., C. Mukasa, A. Tibazalika, E. Mwangi, A.Y. Banana, P. Wamala, and G. Okiror. 2019a. *Strengthening Tenure Security and Community Participation in Forest Management in Kibaale District, Uganda: Insights from Participatory Prospective Analysis*. Bogor, Indonesia: CIFOR. https://www.cifor.org/knowledge/publication/7522/

Mshale, B., A. Tibazalika, C. Mukasa, E. Mwangi, A.Y. Banana, P. Wamala, and G. Okiror. 2019b. *Securing Tenure Rights to Communal Forests in Masindi District, Uganda: Lessons from Participatory Prospective Analysis*. Bogor, Indonesia: CIFOR. https://www.cifor.org/knowledge/publication/7517/

Mukasa, C., A. Tibazalika, P. Wamala, G. Okiror, G. Ngatia, P. Maina, and E. Mwangi. 2019. *Advancing Forest Tenure Security in Nyeri County, Kenya: Learnings from Participatory Prospective Analysis (PPA)*. Bogor, Indonesia: CIFOR. https://www.cifor.org/knowledge/publication/7523/

Mwita, V.C., D. Kalenzi, and F. Flintan. 2017. "Joint Village Land Use Planning in Tanzania: A Process to Enhance the Securing of Rangelands and Resolving Land Use Conflicts." Presentation, 2017 Conference on Land Policy in Africa, Addis Ababa, Ethiopia, November 14–17. https://hdl.handle.net/10568/89927

Ortiz-Guerrero, C., J.H. Arango, V.G. Solórzano, and P.R. Barón. 2017. "Historical Trajectories and Prospective Scenarios for Collective Land Tenure Reforms in Community Forest Areas in Colombia." Working Paper 237. CIFOR, Bogor, Indonesia. https://doi.org/10.17528/cifor/006797

Oyono, P.R. 2021. *Poverty, Tenure Security, and Landscape Governance: Exploring Inextricable Interdependencies for Science, Policy, and Action*. PIM Flagship Brief. Washington, DC: IFPRI. https://doi.org/10.2499/p15738coll2.134791

Pradhan, R., R. Meinzen-Dick, and S. Theis. 2019. "Property Rights, Intersectionality, and Women's Empowerment in Nepal." *Journal of Rural Studies* 70: 26–35. https://doi.org/10.1016/j.jrurstud.2019.05.003

Quan, J., J. Monteiro, and P. Mole. 2013. "The Experience of Mozambique's Community Land Initiative (ITC) in Securing Land Rights and Improving Community Land Use: Practice, Policy and Governance Implications." Presentation, 2013 World Bank Conference on Land and Poverty, Washington, DC, April 8-11. https://www.oicrf.org/-/the-experience-of-mozambique-s-community-land-initiative-itc-in-securing-land-rights-and-improving-community-land-use-practice-policy-and-governance-i

Quisumbing, A.R., and N. Kumar. 2014. *Land Rights Knowledge and Conservation in Rural Ethiopia Mind the Gender Gap*. IFPRI Discussion Paper 01386. Washington, DC: IFPRI. http://ebrary.ifpri.org/cdm/ref/collection/p15738coll2/id/128480

Ranjatson, P., R. McLain, J.M. Rakotonirina, R. Randrianasolo, N.T. Razafimbelo, and S. Lawry. 2019. "Tenure Security and Forest Landscape Restoration: Results from Exploratory Research in Boeny, Madagascar." Presentation, World Bank Conference on Land and Poverty, Washington, DC, March 25–29. https://www.cifor.org/knowledge/publication/7212/

Ratner, B., C. Burnley, S. Mugisha, E. Madzudzo, I. Oeur, K. Mam, L. Rüttinger, L.N. Chilufya, and P. Adriázola. 2017a. "Facilitating Multi-stakeholder Dialogue to Manage Natural Resource Competition: A Synthesis of Lessons from Uganda, Zambia, and Cambodia." *International Journal of the Commons* 11 (2): 733–753. https://doi.org/10.18352/ijc.748

Ratner, B., R. Meinzen-Dick, J. Hellin, E. Mapedza, J.D. Unruh, W. Veening, E. Haglund, C. May, and C. Bruch. 2017b. "Addressing Conflict through Collective Action in Natural Resource Management." *International Journal of the Commons* 11 (2): 877–906. http://doi.org/10.18352/ijc.768

Ratner, B., C. Burnley, S. Mugisha, E. Madzudzo, I. Oeur, K. Mam, L. Rüttinger, L. Chilufya, and P. Adriázola. 2018. "Investing in Multi-Stakeholder Dialogue to Address Natural Resource Competition and Conflict." *Development in Practice* 28 (6): 799–812. https://doi.org/10.1080/09614524.2018.1478950

Ratner, B.D., A. Rivera, and A. Fiorenza. 2019. "Engaging Government for Policy Influence through Multi-stakeholder Platforms." Guidance Note. Collaborating for Resilience and International Land Coalition, Rome. https://www.landcoalition.org/en/resources/engaging-policy-makers-influence-policy-guidance-note/

Resnick, D. 2021. *Embracing Political Economy to Enhance Policy Influence: Lessons from PIM Research*. PIM Synthesis Brief June 2021. Washington, DC: IFPRI. https://doi.org/10.2499/p15738coll2.134440

Robinson, B.E., Y.J. Masuda, A. Kelly, M.B. Holland, C. Bedford, M. Childress, D. Fletschner, et al. 2018. "Incorporating Land Tenure Security into Conservation." *Conservation Letters* 11 (2): 1–12. https://doi.org/10.1111/conl.12383

Schlager, E., and E. Ostrom. 1992. "Property-Rights Regimes and Natural Resources: A Conceptual Analysis." *Land Economics* 68 (3): 249–262. https://doi.org/10.2307/3146375

Sharma, B.P., S. Lawry, N.S. Paudel, R. McLain, A. Adhikary, and M.R. Banjade. 2020. "Operationalizing a Framework for Assessing the Enabling Environment for Community Forest Enterprises: A Case Study from Nepal." *Small-scale Forestry* 19: 83–106. https://doi.org/10.1007/s11842-020-09433-w

Shelton, R.E., M.A. Janssen, and R. Meinzen-Dick. 2018. "Measuring Learning from Interventions through Participatory Processes." CBIE Working Paper Series CBIE-2018-004. Center for Behavior, Institutions and the Environment, Tempe, AZ. https://cbie.asu.edu/sites/default/files/papers/cbie_wp_2018-004.pdf

Stoian, D., and A. Rodas. 2018. "Successful Community Stewardship of Tropical Forests: Evidence from Community Forest Concessions in Petén, Guatemala." Presentation at the 19th Annual Conference on Land and Poverty, World Bank, Washington, DC, March 19–23. https://hdl.handle.net/10568/93439

Stoian, D., A. Rodas, M. Butler, I. Monterroso, and B. Hodgdon. 2018. *Forest Concessions in Petén, Guatemala: A Systematic Analysis of the Socioeconomic Performance of Community Enterprises in the Maya Biosphere Reserve*. CIFOR Brief 7163. Bogor, Indonesia: CIFOR. https://www.cifor.org/knowledge/publication/7163/

Suhardiman, D., J.S. Bright, and C. Palmano. 2019. "The Politics of Legal Pluralism in the Shaping of Spatial Power in Myanmar's Land Governance." *Journal of Peasant Studies* 48 (2): 411–435. https://doi.org/10.1080/03066150.2019.1656200

Suhardiman, D., O. Keovilignavong, and M. Kenney-Lazar. 2019. "The Territorial Politics of Land Use Planning in Laos." *Land Use Policy* 83: 346–356. https://doi.org/10.1016/j.landusepol.2019.02.017

Sulle, E., and W. Mkama. 2019. *A Contextual Analysis for Village Land Use Planning in Tanzania's Bagamoyo and Chalinze Districts, Pwani Region and Mvomero and Kilosa Districts, Morogoro Region*. ILRI Project Report. Nairobi, Kenya: ILRI. https://hdl.handle.net/10568/105443

Tibalazika, A., C. Mukasa, P. Wamala, G. Okiror, E. Mwangi, M. Buloa, and D. Ojwang. 2019. *Enhancing Forest Landscape Restoration in Mount Elgon, Kenya: Insights from Participatory Prospective Analysis (PPA)*. Bogor, Indonesia: CIFOR. https://www.cifor.org/knowledge/publication/7521/

UNDESA (United Nations, Department of Economic and Social Affairs, Population Division). 2019. *World Urbanization Prospects: The 2018 Revision*. New York: United Nations. https://esa.un.org/unpd/wup/

Valkonen, A.H. 2018. "Conceptions, Practices and Power Plays around the Notion of Tenure Security: Examining the Land Policy Development and Implementation Process in Madagascar." PhD Dissertation, School of International Development, University of East Anglia, UK. https://ueaeprints.uea.ac.uk/id/eprint/70661

Velasquez-Ruiz, M.A. 2018. *Collective Land Tenure in Colombia: Background and Current Status*. CIFOR Infobrief 217. Bogor, Indonesia: CIFOR. https://doi.org/10.17528/cifor/006902

Werner, J., A. Nefzaoui, J. Kailene, and M. Louhaichi. 2018. *A New Pastoral Code for Tunisia: Reversing Degradation Across the Country's Critical Rangelands*. International Center for Agricultural Research in the Dry Areas (ICARDA) Policy Brief. Beirut, Lebanon: ICARDA. https://hdl.handle.net/20.500.11766/8806

Zamora, A., and I. Monterroso. 2019. *Regional and Local Perspectives on Tenure Insecurity in the Loreto and Madre de Dios Regions of Peru*. CIFOR Infobrief 246. Bogor, Indonesia: CIFOR. https://doi.org/10.17528/cifor/007235

Made in United States
Cleveland, OH
05 November 2026

47664969R00026